Pvt. Ltd.

Encyclopedia of Healing Plants

Md. Rageeb Md. Usman
Sufiyan Ahmad
Sunil P. Pawar
V. M. Shastry
Mohammed Zuber Shaikh

2017

Studium Press (India) Pvt. Ltd.

Encyclopedia of Healing Plants

ISBN: 978-93-80012-91-9

Published by:

Studium Press (India) Pvt. Ltd.
4735/22, 2nd Floor, Prakash Deep Building
(Near Delhi Medical Association),
Ansari Road, Darya Ganj, New Delhi-110 002
Tel.: + 91-11-43240200-15 (15 lines); Fax: 91-11-43240215
E-mail: studiumpress@gmail.com

Printed at:

Printed In India

Tapi Valley Education Society's
Hon'ble Loksevak Madhukarrao Chaudhari
COLLEGE OF PHARMACY
(Approved by AICTE & PCI
New Delhi & Affiliated by NMU Jalgaon.)
Neharu Vidyanagar, Faizpur (425503)
Dist. Jalgaon
Ph- 02585-245574, Resi-02585-245745
Fax No- 02585-245574
E-mail-copharmafaizpurr@yahoo.com
Website-copfaizpur.org

Prof. (Dr.) Vijay R. Patil
Principal
Member of Management Council
North Maharashtra University, Jalgaon
Dean: Medicine & Pharmacy Faculty
North Maharashtra University, Jalgaon

Foreword

I am really delighted to foreword Encyclopedia of Healing Plants written by authors.

Medicinal Plants has always been major part of Science curriculum. This book covers in detail all the aspect of Medicinal Plants required for studies.

The book contains General Introduction, Plants for Anti-inflammatory, Antiulcer, Antifertility, Hepatoprotective, Antidiabetic, Antipyretic, Anticancer, Antimicrobial and Diuretic in very simplified language to make better understanding of the students.

I congratulate and my best wishes to authors for their sincere efforts for the benefit of the students of Pharmacy Profession.

Prof. (Dr.) Vijay R. Patil
Dean
Pharmaceutical Science
North Maharashtra University
Jalgaon, MH, India

About the Authors

Mohammed Rageeb Mohammed Usman

Mohammed Rageeb Mohammed Usman is Assistant Professor in Pharmacognosy Department at Smt. S.S. Patil College of Pharmacy, Chopda. He has 8 years research and teaching experience. His Publications comprise 2 Reference, 1 Text, 3 Practical & 1 Competitive International books, Published 1 Chapter in International book and more than 55 Review/Research articles in various peer reviewed national and international journals having good impact factor. He has Published General 4 articles in Scientific Update. He also Presented more than 140 oral and poster presentations and participated more than 175 national and international scientific occasions. He is a life time member of various Pharmacy professional Associations (APTI, APP, SPER, ISP, ICPHS, IHPA, IPA, RSH, ABAP, IGPA INHS and IPA). His biography has been included in the renowned directory "Who's Who in the Worldin Vol. XIV (2016). He is President of Society of Pharmaceutical Education and Research (SPER) Maharashtra State Branch, President of Association of Pharmacy Professional Maharashtra State Branch (APP), Indian Pharmacist Association (IPA) and Research Scholar Hub (RSH) Maharashtra State Branch, Society for Researchers and Healthcare Professional (SRHCP), Associate Editor of Tyagi Pharmacy Association (TPA), Associate Editor of SPER Time, Editor in Chief of Annual College Magazine "PHYSIC-2009, PHYSIC-2014 & PHYSIC-2015". He is elected as a Fellowship award of Indian Congress of Pharmacy and Health Science (ICPHS), Association of Pharmacy Professional (APP), Society of Researcher and Health Care Professional (SRHCP), Research Scholar Hub (RHS), Society of Pharmaceutical Education and Research (SPER) and editorial board member of more than 70 Journals and Reviewer/ Referee of more than 21 peer reviewed journals. He has supervised 24 (04 Ongoing; 20 completed) UG Project and extensive work for Pharmaceutical Science. He organized 6 National Conferences, 3 National Seminar, 2 National Health awareness camp and 2 International Conference. He is Guest of Honors, Co-Chair Person, Speaker and Judge for Scientific Session for Various National/International Conferences. He has Received Appreciation/Best Oral/Poster Presentation Award in Various Conferences. Prof. Md. Rageeb Md. Usman conferred with Various Awards Like "Young Performer Award", "Young Achiever Award" and "Appreciation Awards" in March 2013, "Young Pharmacy Teacher Award" in January 2014, "Young Innovative Researcher Award" in March 2014, "Young Talent Award 2014" Oct. 2014, "Young Pharmacist Award 2015" and "Best State Branch Award 2015"

in 31 Jan. 2015, "Best State Branch Award 2015" (SPER Maharashtra State) in February 2015, "Young Academic Excellence Award" in January 2016, "Life Time Achievement Award" in January 2016 and "Young Talent Award" in March 2016.

Dr. Sufiyan Ahmad Raees Ahmad

Dr. Sufiyan Ahmad Raees Ahmad had completed B. Pharm from A.R.A. College of Pharmacy, Dhule (North Maharashtra University), M. Pharm from Vinayaka Mission's College of Pharmacy, Salem (T.N.) and Doctorate in Pharmacy from JJT University (Rajasthan). He is presently working as associate professor and head of department in Pharmacognosy at Gangamai College of Pharmacy, Nagaon, Dist. Dhule, Maharashtra, India. He is life member of various Professional organizations such as Association of Pharmaceutical Teachers of India (APTI), In Pharm Association 'A Young Pharmacist's group, Association of Pharmacy Professionals (APP) and Society of Pharmaceutical Education and Research (SPER). He has guided 6 PG students. UGC has sanctioned one minor research project to him. He has participated in various National and International conferences. He has extensive work in the field of Pharmaceutical Sciences and has contributed number of research and review article publications in high reputed international and national journals.

Prof. Dr. S.P. Pawar

Prof. Dr. S.P. Pawar has nearly 18 years of research and teaching experience at both U.G and P.G levels. He is a leading scientist and well-known academicians of international repute. He did his Ph.D under the supervision of Prof. Dr. P. G. Yeole (Pro-VC, Nagpur University) at the department of pharmaceutical science, Nagpur University. He has supervised 08 Ph.D (06 Registered; 02 completed) and more than 05 M. Pharm research Projects. More than 55 research publications are on his credit published in international and national journals. About 47 research abstracts has been presented by him in various proceeding of national and international conference. He is a widely travelled person in India as inspection committee member and chairman, appointed by Pharmacy Council of India and North Maharashtra University, Jalgaon. He takes very active interest in professional matters and under his leadership and administration college of pharmacy, Shahadawas accredited by NBA. He has been honored by NMU Jalgaon as best administrator in pharmacy division and appointed as chairman, Board of studies for subject's Pharmacognosy and Pharmacology. Dr. S. P. Pawar was appointed as regional co-ordinator of North Maharashtra region, by Association of Pharmaceutical teachers of India (Maharashtra state branch). He was also served as a joint a secretary for 10th APTI National convention, 2005 organized by department of pharmaceutical science, Rashtrasant Tukdoji Maharaj, Nagpur University, Nagpur. He has delivered invited lectures, keynote addresses and chaired many session in several national and international conferences as well as symposia in India. Presently he is serving as principal, P.S.G.V.P. Mandal's college of pharmacy, Shahada affiliated to North Maharashtra University, Jalgaon.

Prof. (Dr.) Vijay M. Shastry

Prof. (Dr.) Vijay M. Shastry had completed B. Pharm & M. Pharm from MVP's College of Pharmacy, Nashik (Pune University). He pursued Doctorate in Pharmacy from North Maharashtra University, Jalgaon (MS). Pharmacy profession is his religion and academics is his passion. Presently he is holding the post of Professor in Pharmacognosy & Principal at Nagaon Education Society's Gangamai College of Pharmacy, Nagaon, Dhule (North Maharashtra University). Number of research and review articles are published in national and international journals on his credit. He has also presented papers in several conferences.

Dr. Mohammed Zuber Shaikh Usman

Dr. Mohammed Zuber Shaikh Usman has been awarded Ph.D degree from University of Mumbai, Mumbai. He started his carrier as a Senior Research Fellow in one of the best ICAR research Centre located in Mumbai on World Bank Funded project titled as 'Climate Change Adaptation'. He has worked as assistant professor in Sathaye College, Vile Parle (E), Mumbai. He has Presented/Participated and published many Research and Review articles in peer reviewer international journal and conferences, even he hold chairmanship in scientific sessions of many conferences. He published one International book. He guided under graduate students on various research base project. Presently he is working as Head of Zoology Department, Senior Science College, Akkalkuwa, Maharashtra affiliated to North Maharashtra University, Jalgaon. He is selected as a Vice President of Society of Pharmaceutical Education and Research (SPER) & Association of Pharmacy Professional (APP) and Honor as a Life Membership of SPER & APP. He is also Expert for Poster Evaluation in various National and International Conference & Exhibition.

Preface

This book having title "Encyclopedia of Healing Plants"written with an intention to benefit the Diploma, UG, PG and Research Scholar students.

The generally acknowledged importance of study of Medicinal Plants as a useful knowledge and the constantly increasing recognition of its extended practical as well as scientific applications.

Since time immemorial, human beings have learned on plants/herbs/shrubs etc. for curative treatment of diseases and to secure prevention and cure against manifestations of various ailments.

We have felt the urge of compilation of a Handbook on medicinal plants, with up-to-date scientific information and educative/discovered, formulae, after being persuaded by the requests from Pharmaceutical researchers for a comprehensive compilation on this subject.

The aim of this compilation is to highlight information on scientific Nomenclature Local names, Distribution, Parts used. Pharmacological activities and Chemical Constituents of the different medicinal plants, meant for use medicinal purposes. The information collected from different Books, Scientific Journals etc.

The materials and data compiled in this book may be of use by a variety of professionals including Pharmacists, Pharmacologists, Medicinal chemists, Toxicologists, Pharmacognosists, Botanists etc.

We and the Publisher cannot be urged to accept and responsibility in the event of any error or omission creeping in or in compilations or printing, keeping in view, the human errors as well as the ever-changing scientific advancement of knowledge all over the world by continuous penetrating researches, according to developments, gradually coming up to surface rapidly.

Authors

Acknowledgement

The authors are thankful to Prof. Dr. V. R. Patil, Dean of Pharmaceutical Science, North Maharashtra University, Jalgaon, Maharashtra, India and Principal, H'ble LMC's TVES's. College of Pharmacy, Faizpur, Member of Management Council, NMU, Jalgaon for his valuable guidance and critical suggestions.

The authors are grateful to Dr. Mohib Khan, Professor, Oriental College of Pharmacy, Sanpada, Navi Mumbai for their appreciation, moral support, constant encouragement, positive criticism and scientific inputs.

We are thankful to the Management, Nagaon Education Society's Gangamai college of Pharmacy, Nagaon, Dist. Dhule, Maharashtra, India, for providing the necessary infrastructure and other facilities.

We are thankful to the Management and Prof. Dr. G. P. Vadnere, (Principal) Smt. Sharadchandrika Suresh Patil College of Pharmacy, Chopda, Maharashtra, India for excellent guidance and dedicated efforts.

We are grateful to our parents for their unconditional love, support and encouragement.

The authors welcome suggestions and comments for improvement of the book.

We are thankful to publishers and extend our thanks to supportive friends, colleagues and for bringing out nicely printed book.

Authors

Table of Contents

1

GENERAL INTRODUCTION

INTRODUCTION

Since the origin of man on earth, man has been using plants and herbs as a food and then as a medicine. In 19th century research has been developed and life saving drugs has been isolated from these plants and herbs. Because of more side effects, high cost of modern medicines, 80% of world's population including western countries is turning towards herbal medicines. It is because of availability, safety, in accessibility to neigh boring town and mainly affordability the population in village prefer using herbal drugs.

Though, there has been a tremendous growth of the traditional systems of health care worldwide, recently in countries like India, China and Brazil health care scenario has always been associated with the traditional system of medicine. These countries are still having very rich biological as well as cultural diversity and the traditional health care systems have a deep scenario in these nations. It has been estimated that from 25000 to 75000 species of higher plant, about 10% has been used in traditional medicine. However perhaps only about 1% of these (250-750) are acknowledge through scientific studies to have therapeutic value when used in extract form by human.

Plant derived medicines are useful therapeutic options and often provide a safe form of therapy, and in many instances specific phytomedicines have been shown to be clinically effective. There are many reasons for increased use of herbal medicines. These may range from appeal of product from 'nature' and the perception that such product are 'safe' (or at least 'safer than conventional medicines, which are often referred to as 'drugs') to more complex reasons related to philosophical views and religious beliefs of individuals.

Based on the Traditional System of medicine, the use of a crude plant for a particular therapeutic activity, its extraction and fractionation may be carried out. The fractions of extract may prove to be effective, and less toxic for a particular ailment, when compared to an already existing alternative drug.

Nature has provided a complete store house of remedies to cure all ailments of mankind. Since the down of civilization, in addition to food crops, man cultivated herbs for his medicinal needs. The knowledge of drugs has accumulated over thousands of years as a result of man's inquisitive nature, so that today we possess many effective means of ensuring health care. The history of herbal medicines is as old as human civilization. The documents, many of which are of great antiquity, revealed that plants were used medicinally in China, India, Egypt and Greece long before the beginning of the Christian era.

In Indian system of medicine, a large number of drugs of either herbal or mineral origin have been advocated for various types of diseases and other different unwanted conditions in humans. Ayurveda is one of the traditional systems of medicine practiced in India and Sri Lanka and can be traced back to 6000 BC. Ayurvedic medicines are largely based upon herbal and herbomineral preparations and have specific diagnostic and therapeutic principles.

Plant-derived medicines are useful therapeutic options and often a safe form of therapy and in many instances specific phytomedicines have been shown to be clinically effective. There are many reasons for increased use of herbal medicines. These may range from appeal of product from 'nature' and the perception that such product are 'safe' (or at least 'safer than conventional medicines, which are often derogatorily referred to as 'drugs'), to more complex reasons related to philosophical views and religious beliefs of individuals.

Medicinal plants are accessible, affordable and appropriate source of primary healthcare for almost two-third population of our country. People who can't afford high cost of modern medicines are mainly dependant on these culturally familiar, simple, affordable and effective traditional medicines. Because of these favorable factors, there is wide spread interest in promoting traditional health system to meet primary health care needs. The spreading cost of modern medicines has made it necessary for the protection and promotion of the cultural and spiritual values of traditional medicines.

Ayurveda an ancient system of Indian medicine has recommended in number of drugs from indigenous plant/animal sources for the treatment of several diseases or disorders.

The use of plants for treating various diseases predates human history and forms the origin of much of the modern medicine. Long before the advent of modern medicine, herbs were the mainstream remedies for nearly all ailments. People commonly diagnosed their own illnesses, prepared and prescribed their own herbal medicines or brought them from the local apothecaries. According to World Health Organization, herbal medicine is defined as plant derived material or preparation, which contains raw or processed ingredients from one or more plants with therapeutic values. Use of plant products, as medicine is inherent in Ayurveda, the ancient Indian system of health care. The use of medicinal plant is accepted as the most common form of traditional medicine. Among the entire flora, it is estimated that 35,000 to 70,000 species have been used for medicinal purpose. Some 5000 of these have been studied in biomedical research. In developing countries, herbal medicines continue to play important role in primary health care, especially where coverage of health service is limited.

In industrialized countries, herbal medicine are become increase in popular, However the expanded use of herbal medicine has lead to concerns relating to assurance of safety, quality and efficacy.

An acute interest in the phytochemistry and pharmacology of plant drugs is evident from the number of research paper published in different journals and through finding presented in national and international conferences. The thrust of the herbal drug research may be due to different factors one may be that no major new leads are covering up from research on synthetic therapeutic agents and antibiotics.

Herbs and plants are valuable not only for their active ingredients but also for their minerals, vitamins, volatile oils, glycosides, alkaloids, acids, alcohol, and esters and these component come from all parts of the plant including leaves, flowers, stem, berries, seeds, fruit, bark and roots.

Synthetic drug research has yielded number of medicine for the treatment of various elements like tuberculosis, epilepsy, gastric ulcer, hypertension, anxiety and Parkinsonism etc. but at the same time there are many diseases for which there is no definite cure available in modern system of medicine. In this situation we could look to the plant kingdom for remedy. Plants have provided the drugs like digoxin, vincristine, vinblastine or quinine on the one hand and raw material like diosgenin or solasodine for the synthesis of anabolic, estrogenic and androgenic steroid on the other hand.

Thus plant drug research appears to be complementary to the ongoing synthetic drug research. (World Health Organization in its technical report on promotion of development of traditional medicine).

India has rich heritage of usage of medicinal plant in its Ayurvedic and Unani system of medicine besides use of many plant in folk remedies. The major populations of the country particular in ruler/tribal areas rely heavily on the use of herbal medicine for treatment of various diseases. While India adopted the allopathic system medicine in its cities and towns, the indigenous system of medicine survived in the ruler area due to Government or other related agencies support these factor ensured the continued presence and usage of herbal products of the Indian scene. An acute interest in the phytochemistry and pharmacology of plant drugs is evident from the number of research paper published in different journals and through finding presented in national and international conferences. The thrust of the herbal drug research may be due to different factors one may be that no major new leads are covering up from research on synthetic therapeutic agents and antibiotics.

Secondly many of the antibiotics and synthetic drugs have shown sensitization reaction and other undesirable side effects and there is feeling that the herbal drug are comparatively safe.

Since the early human existence, many natural products came into practice for human welfare by sheer intuition or more appropriately by trial and error. Because of this practice, in fact, every country including the civilization of China, Egypt and India developed its own medical system. Thus, the Indian medical system Ayurveda, came into existence. The raw materials used for ayurvedic medicines' were dried herbal powders or their extracts or mixture of products.

In the long struggle to overcome the powerful forces of nature, the human beings have always turned to plants for food, shelter, clothing, weapons and healing. When pain or injury or diseases struck, ancient people had little choice but to turn to plants. Developed empirically, by trial and error, May herbal preparations were remarkably effective. Then came the modern concept of medical science herbal medicine was left behind, ignored and mandated as superstition. With the scientific age extending its horizon by leaps and bounds, misconceptions on myths persisted. One such myth says **"synthetic is the best"**. If a medicine does not contain purified chemicals, manufactured with sophisticated and expensive equipment, highly trained scientists, the myth says possibly such medicine cannot work. Scientist's now-a-days carry out extensive tests and clinical trial before releasing a drug in the market. But unfortunately there is still room for tragic errors-the one that occurred with Thalidomide.

Throughout human existence, plants were virtually all that was available to healers. This fact is true even today, of course, outside the developed world. Even though synthetic medicine has taken a front seat, 25 to 30 percent of modern drugs are derived from some parts of higher plants. In most cases, application of modern medicine resembles the traditional use of the plant from which it is obtained. The modern era has seen some decline in the use of medicinal plants or their extract as therapeutic agent, particularly in developed countries. Modern medicines and herbal medicines are complimentarily being used in areas for health care programme in several developing countries including India. Of late, the interest in the plant products surfaced all over the world due to the belief that many herbal medical are known to be free from side effects. That apart, the fact that the discovery of the new synthetic drug is time consuming, expensive affair; and if we want to implement "Health for all by the year 2000", then, perhaps the logical choice would lie in exploring plant products as potential medicines to the maximum possible usage.

Plants have provided mankind a large variety of potent drugs to alleviate suffering from diseases. In spite of spectacular advances in synthetic drugs in recent years, some of the drugs of plant origin have still retained their importance. The use of plant-based drugs in the Western world is increasing.

Our ancestors believed 'sickness' to be punishment from the Gods. In early medicines, people were treated with prayers, rituals and magic potions (medicinal preparation from herbs).

Egyptian civilization has given the world-one of its first medical texts the Ebers Papyrus. The papyrus, claimed to have been written sometimes in the 16th century B.C., contains nearly 800 recipes and over 700 references and drugs, which include aloe, Replace by worm wood, peppermint, myrrh, hump dog leave, castor oil. *Ebers Papyrus* also mention Egyptian recipe for treatment of diabetes.

About 2000 years ago, the Chinese pharmacopoeia, the Pen Tsao appeared. Attributed to emperor *Shen Wung,* among many references, the *Pen Tsao* described the use of chaulmoogra oil from *Hydnocarpus* for treatment of leprosy. Ancient Chinese records refer the use of desert Shrub or Chinese Ephedra (ma huang) in fevers, to suppress coughs, relieve lungs and bronchial disorders.

In India, many generations of medical treatment has been recorded in the Ayurveda, one of the earliest collections of Hindu medical lore. The doctrine itself dates back much earlier, to the Rig Veda (references on 67 herbal drugs can be found). Charaka Samhita (900 B.C.) is the first recorded treatise on Ayurveda, which describes 341 plants and plant products for use in medicine. Sushruta Samhita (600 B.C.) records many more herbal remedies and even laid special emphases on surgery. Bhava Mishra's Bhava Prakash (1550 AD) is one of the last celebrated treatise on Hindu system of medicine.

In ancient Greece, medicine was practiced by priests, better known as *Asclepiads* (from Asclepius the Greek God of healing). Treatment was a religious ritual, involving incantations and mystery. Around 400 BC, Hippocrates brought a more scientific approach to the Greek medical treatment. He stressed on the fact that medicine was a science and an art – this infact had earned him the position of 'Father' of Modern Medicine. Theophrastus (Aristotle's pupil) a botanist, whose treatise 'Inquiry into Plants' is a famous work which had a long standing influence both on botany and medicine. In the first century A.D., Dioscorides performed a pioneering work to compile De Materia Medica an authoritative text on the use of various medicinal plants – the guide to all modern/pharmacopoeia.

Scientific curiosity awakened during the Renaissance. The pace of medical discovery has picked up dramatically by the 19th century. Alongside this burst of scientific and technological discoveries, the story of plants as medicines had however continued to unfold.

Inspite of the tremendous advances made in the modern medicine (post Second World War era), there are still a large number of ailments for which suitable drugs are yet to be found. Today, there is an urgent need to develop safer drugs for the treatment of inflammatory disorders, liver diseases, gastrointestinal disorders (ulceration) and last but not the least the problem of population explosion-suitable and safe anti-fertility agents which will have lesser undesirable side effects. Medical science has developed tremendously with remarkable progress in various fields. In this era of specializations, human beings are seen as chemo-physical conglomeration of many separate parts, thus giving rise to new trends of biological thinking.

Even though profound changes are taking place in the medical world, the modern drug approach has a limited arsenal to combat the meteoric increase in deadly disorders such as inflammatory conditions and arthritis. Practitioners of traditional medicinal systems, in fact, do possess cures for ailments such as arthritis, are however branded as 'quacks' Little progress has been made to achieve permanent cure for arthritic and inflammatory disorders. Inflammatory diseases including rheumatism are very common throughout the world. This is one of the oldest known diseases of mankind and affects a large population of the world. The demand of anti-rheumatic drugs will constantly increase as our life expectancy is increasing. So-called systematic study on anti-inflammatory effects of Indian medicinal plants was done by many investigating workers. About 40 plants have been identified as having definite anti-inflammatory activity.

In fact the presently used synthetic drugs are potentially toxic and symptoms reappear following their discontinuation. Thus arthritis today, has crippled thousands, may be millions of men and women, around the world. However, with several ancient civilizations, as that of ours, there exist herbal remedies, which have been in use for several thousands of years for treatment of arthritic conditions. These herbs are not used empirically. In fact, practitioners of traditional medicines in India, Nepal, Sri Lanka, Bangladesh may also provide the rationale behind the use of such herbs. If we look back to the history of Aspirin (discovered in 1899, being one of the most widely used drugs even today), we find that it owes its origin to salicylic acid, which was widely used in ancient folk medicine. The word Salicylate is it self derived from Salicaceae, the botanical name for the willow family. The bark of the white willow trees (*Salix alba*) contains a glycoside (salicin), which possess antipyretic and analgesic properties. Salicin on hydrolysis produces salicylic acid and glucose. The nature and composition of Salicylic acid was established by Kolbe in the middle of 19th century. Thus there is a definite prospect of finding suitable anti-inflammatory and anti-rheumatic drugs from indigenous plants. A systematic study of screening of medicinal plants for anti-inflammatory activity was initiated by Gujrat *et al.* Subsequently various researchers from different parts of India have reported anti-inflammatory activity of many more plants. At the Central Drug Research Institute (CDRL) Lucknow and Regional Research Laboratory (RRL) Jammu, intensive studies are being made for evaluation of anti-inflammatory activity of different indigenous plants.

Disturbances of the gastrointestinal function are responsible for various illnesses and discomfort in human beings. Over the past few decades, there has been a surge in research activity, aimed towards development of effective and safer antiulcer drugs both synthetically and from natural resources. In the 1960's investigations with the roots and rhizomes of liquorice (*Glycyrrhiza glabra*) showed that the plant showed great promise in treatment of peptic ulcer. Doll *et al.* reported their findings on clinical trial of Triterpenoids liquorice compound as antiulcer agents. Since then, a considerable amount of research has been done on Glycyrrhizin, thus leading to the development of carbenoxolone sodium. Similar studies with other indigenous plants have lead to the identification of several Indian herbs with significant antiulcer activity.

The name 'diabetes' is derived from the Greek for syphon, was given to the disease by Aretaeus of Cappadocia (81-138 A.D.). The adjective 'mellitus' owes its origin to Latin for honey, was added by Thomas Willis in 1674. In 1927 the active anti-diabetic principle (insulin) was prepared. The year 1960 is a landmark in history of protein chemistry, when Sanger and his colleagues established the structure of insulin. At present several million people are suffering from diabetes (a potentially dangerous degenerative disease) in India. Herbal remedies for diabetes have been recorded in ancient medical literatures. In the last few decades, the search for newer anti-diabetics from natural resources has intensified. Handa *et al.* have reviewed 148 plant belonging to 50 different families reported to possess hypoglycemic activity. Researchers from around the globe have reported the importance of diet, in the management of diabetes mellitus. Incorporation of fibrous vegetable, fenugreek seeds into the diet of diabetic patients have resulted in significant reduction of blood glucose levels. Plants hold definite promise in the management of Diabetes mellitus.

Liver disorders is one the most common problems countenanced in a country like India, since liver being one of the most vital organs of the human system, performing diverse function, a healthy liver is one of the prerequisites for a normal and healthy life. If we take a look at the medical statistics, we find, there are about 20,000 deaths globally due to cirrhosis of liver. Liver carcinoma is one of the most common forms of cancer which is prevalent in the world today year. Today, there is an increasing trend in liver disorders and this may be attributed to an alarming increase in pollution levels, increased consumption of alcohol, Indiscriminate and irrational use of drugs, particularly antibiotics; and of course, due to viral infections. Modern system of medicine has a very limited application in alleviation of hepatic disorders but we have in numerable reports where herbal medicines have worked wonders in curing liver disorders. A statistical analysis of prescriptions through out India for liver disorders might indicate a growing trend of prescribing herbal medicines. Thus a careful and extensive study of medicinal plants is necessary to identify newer plant products, isolation of purified compounds and this in turn may help us to combat the dreaded problem of liver disorders. Until recently, it has been accepted almost as dogma that there is none and that there cannot be any pharmacological treatment for liver disease. Only drugs used were cortico steroids immuno suppressive agents. Although a number of plants were used in Ayurveda for liver diseases, till the work on Silybum *marianum* was published by western workers, our scientists were not so alive on the subject. Now more than 25 plants of Indian origin have been investigated and confirmed to possess these properties. When we take a look at a country like India, we can easily understand that indiscriminate and uncontrolled growth of human population in our country, is one of the biggest problems that we are facing today population explosion is a problem which has assumed a global dimension. With a meteoric increase in human population, we have started facing problems related to scarcity of water, shelter, food. There is indiscriminate deforestation, conversion of forest areas into agricultural land for increased generation of food grains thus leading to serious ecological disturbances. The whole world feels that immediate steps are necessary to curb such uncontrolled growth of human population. The Government of India has formulated National policies to reduce the growth rate but alas! We have achieved very little. Along with conventional antifertility devices available in the market, uses of steroidal hormones are also being prescribed by doctors to control birth. Owing to the undesirable side effects of the available hormonal preparations, it is necessary that we have some relatively non toxic drugs, which will have low side effects on long term use. Over the ages, plant preparations have been used effectively by practitioners of traditional medicine to control birth. On survey of medical literature, numerous reports are available, where plant preparations have been found to be effective as anti-fertility agents with an advantage of low side effects. Thus herbal preparations can provide us with an alternative in controlling the global problem of population explosion.

The last few decades have seen a different approach by scientists and medical practitioners where herbal remedies are concerned. In fact, botanical medicine has made a come back. Even though modern science has disposed off the fanciful and emotional claims of herbal medicines, it has definitely acknowledged the herbal treatments and plant medicines which worked. We have not been able to replace digitalis or quinine with a better synthetic drug. In fact, new medical science is reaffirming such of the old herbal remedies and thus extending the horizons of botanical medicine.

It is worth while and pertinent to mention that plants may offer a medical revolution to several million people of the world who primarily rely on traditional medicine for their health care needs, in particular, the third world countries. According to WHO estimates, 80% of about 4000 million inhabitants primarily depend on plant products. According to a British survey estimate, about 5000 million sterling pound equivalents of plant products are sold in the developed countries alone. Recently the WHO scientists have come up with a novel idea on 'Alternative medicines' (remedy with plants) for the developing nations, where modern medicines are prohibitively expensive and advanced medical technology is in the developmental stages. Scientists indicate, such nations could in fact raise their own medicine, from plants which already exist around their respective home lands.

As we approach the 21st century, modern technology and scientific discoveries are ushering remarkable changes in our lives, never the less, the story of plants as medicines definitely continues to unfold, however, quietly and independently.

HERBAL

Drugs came into existence since a very early time to remove the pain of diseases and to cure them. Drugs used in medicine today are either obtained from nature or are of synthetic origin. Thus, the story and the history of drugs are as old as the mankind.

Herbal Medicine is defined as a branch of science in which plant based formulations are used to palliate diseases. It is also known as botanical medicine or phytomedicine. Lately phytotherapy has been introduced as more accurate synonym of herbal or botanical medicine. In the early twentieth century herbal medicine was pre-eminent healthcare system as antibiotics or analgesics were not as yet discovered. With the emergence of allopathic system of medicine, herbal medicine gradually lost its popularity among people, which is based on the fast therapeutic actions of synthetic drugs.

Return to Nature: Recently there has been a parallel in and out trend from synthetic to herbal medicine. Medicinal plants have been known for millennium as they are highly a rich source of therapeutic agents for the prevention of diseases and ailments. The search for eternal health and longevity and for remedies to relieve pain and discomfort drove early man to explore his immediate natural surroundings and led to the use of many plants, animal products, minerals etc. and the development of a variety of therapeutic agents.

Plants have been used as medicines throughout history. Studies of wild animals show that they also intuitively eat certain plants to treat themselves for certain illnesses. The practice of herbal medicine is extremely well established and documented in Asia. As a result, most of the medicinal plants that have international recognition come from China and India. In Europe and North America, the use of herbal medicine is increasing fast, especially for correcting imbalances caused by modern Fast food diets and lifestyles. Now medicinal plant products have been taken by many people on a daily basis, to maintain good health as much as to treat diseases. The world is gaining increasing attention towards the importance of medicinal plants and traditional health systems in solving the health care problems. As from virtual extinction of interest, the research on plants of medicinal importance is growing phenomenally, often to the disservice of natural habitats and mother populations in the countries of origin. As an integral part of their culture most of the developing countries have adopted traditional medical practice.

Historically, all medicinal preparations are obtained from plants, whether in the simple form of raw plant materials or in form of crude extracts, mixtures, etc. Biologically active compounds from higher plants have played a vital role in providing medicines to struggle against pain and diseases. For example, in The British Pharmacopoeia (1932), over 70% of organic monographs are on plant-derived products. The plant derived from therapeutic agents significantly decreased (mostly) in the economically developed nations, due to the arrival of synthetic medicines, and subsequently of antibiotics. Thus, the share of plant-based monographs fell to approximately 20% in The British Pharmacopoeia (1980). The share of plant-based drugs has been not more than 2% as new chemical entities introduced as medicinal agents over the past several decades. The revivification of interest in plant remedies has been spurred on by several factors.

- The effectiveness of plant medicines
- Source of direct therapeutic agents
- Affordable by the people
- Elaboration of more complex semi-synthetic chemical compounds from raw material base
- Models for new synthetic compounds
- Taxonomic markers for the discovery of new compounds
- The production of medicinal plants and its consumption for international trade. Phytomedicine are expected to grow in future quite significantly.
- Renewable source
- The preference of consumers for natural therapies, a greater interest in alternative medicines and a commonly held belief for superiority of herbal products then manufactured products
- Displeasure with the results from synthetic drugs and the conviction that herbal medicines may be effective in the treatment of certain diseases where predictable therapies and medicines have proven to be inadequate.
- Despite its harmful effect on long term usage.
- Quality control of herbal medicines with recent technique.
- Patients started realizing serious effects of these he looks back into the older life style. So ultimate example for this is herbal usage.
- A movement towards self-medication

Investigation of the chemical and biological activities of plants during the past two centuries have yielded compounds for the development of modern synthetic organic chemistry as a major route for discovery of novel and more effective therapeutic agents.

For the most plant, the discovery of the drugs can be possible from traditional knowledge that plant parts or extracts can be used to treat one or more diseases in humans. The more interesting of the extracts are then subjected to pharmacological and chemical tests to determine the nature of the active components. Therefore, it should be of interest to ascertain just how important plant drugs are used in the form of crude extracts throughout the world. There is a great deal of interest in support for the search for new and useful drugs from higher plants in countries such as the India, People's Republic of China, Japan, USA, and the Federal Republic of Germany. Virtually every country of the world is active in this search to a limited degree. Higher plants have been described as chemical factories which are capable of synthesizing unlimited numbers of highly complex and unusual chemical substances whose structures could escape the imagination of synthetic chemist forever. Considering that many of these unique gene sources may be lost forever through extinction and that plants have a great potential for producing new drugs of great benefit to mankind, some action should be taken to reverse the current apathy in the United States with respect to this potential.

Herbal Medicines

Have been used in medical practice for thousands of years and recognized especially as a valuable and readily available resource for healthcare in East Asian nations.

India is perhaps the largest producer of medicinal herbs and is rightly called as the **"Botanical Garden of the World"**. Medicinal plants have played a significant role in ancient traditional systems of medication in many countries. Traditional medicine using plant extracts continues to provide health coverage for over 80% of the world's population, especially in the developing world. In India, thousands of plant species are known to have medicinal values and the use of different parts of several medicinal plants to cure specific ailments has been in vogue since ancient time. The medicinal value of the plants lies in some active chemical substances called phytochemicals that produce a definite physiological action on the human body. Phytochemicals are divided into two groups, which are primary and secondary constituents according to their functions in plant metabolism. Primary constituents comprise common sugars, amino acids, proteins and chlorophyll while secondary constituents consists of alkaloids, terpenoids, flavonoids, tannins, phenolic compounds etc.

Pharmacovigilance Systems

The traditional herbal medicines and their preparations have been widely used for thousands of years in India and many oriental countries, such as in China, Korea, Japan, etc. However, one of the characteristics of herbal medicinal preparations is that all the herbal medicines, either presenting as single herb or as collections of herbs in composite formula, is extracted with boiling water during the decoction process. This may be the main reason why quality control of herbal drugs is more difficult than that of western drug. As pointed in "General Guidelines for Methodologies on Research and Evaluation of Traditional Medicines". Despite its existence and continued use over many centuries, and its popularity and extensive use during the last decade, traditional medicine has not been officially recognized in most countries. Consequently, education, training and research in this area have not been accorded due attention and support. The quantity and quality of the safety and efficacy data on traditional medicines are far from sufficient to meet the criteria needed to support its use worldwide. The reasons for the lack of research data are due to not only to health care policies, but also to a lack of adequate or accepted research methodology for evaluating traditional medicine". In general, one or two markers or pharmacologically active components in herbs and or herbal mixtures were currently employed for evaluating the quality and authenticity of herbal medicines, in the identification of the single herb or herbal medicinal preparations, and in assessing the quantitative herbal composition of an herbal product. This kind of the determination, however, does not give a complete picture of herbal product, because multiple constituents are usually responsible for its therapeutic effects. These multiple constituents may work 'synergistically' and could hardly be separated into active parts. Moreover, the chemical constituents in component herbs in the Herbal products may vary depending on harvest seasons, plant origins, drying processes and other factors. Thus, it seems to be necessary to determine most of the phyto-constituents of herbal products in order to ensure the reliability and repeatability of pharmacological activity and clinical research, to understand their bioactivities and possible side effects of active compounds and to enhance product quality control.

When two or more herbs are used in formulations, they are known as poly herbal formulations. Some time herbs are combined with mineral preparations also. The herbs often exist in crude state and Ayurveda describes method of purification of toxic herbs. The concept of poly herbalism is peculiar to Ayurveda although it is difficult to explain in term of modern parameters. Single plant based formulations may have better acceptance from quality control and standardization aspects, but still not ample amount of evidence has accumulated to prove concept of standardization of herbal drugs based on single constituent. Researchers in the last century identified and isolated Salicin, a glycoside as active principle. From salicin, salicylic acid and finally aspirin was synthesized. Aspirin is known to cause gastric irritation and hypersensitivity. The plant when used alone does not cause gastric irritation, probably due to the presence of tannins. It can be concluded that polyherbal formulations should not be dismissed only on the basis that they do not withstand modern research. Ayurveda and herbal medicine has roots in medicinal herbs and they have been practiced for centuries. The western system of medicines too uses complex formulations of plant-based raw material and few of these medicinal plants have generated bulk demand in the organized sector of industry and also in the export.

Now characterization of powder vegetable drugs, there are many techniques available like TLC, HPTLC, UV-Visible spectroscopy, HPLC etc. Those techniques are highly costlier and sophisticated, required more attention for maintenance. While microscopy is an important, cheap and handy tool to determine the identity of the used species in order to authenticate genuine powder vegetable drug. However, once the plants have been processed, it is difficult to identify them through macroscopic identification; other means are necessary. A comparative study can be possible in case of closely related species of plants. Only simple microscope is required for this method.

Authentication of powder vegetable drugs is a critical step in the use of these materials for both research purposes and commercial preparations. Microscopic evaluation and comparison of authenticated and unauthenticated samples of powdered plant material is a cost effective and accurate means of identifying herbal ingredients. Microscopy and computer can be useful tools for the detection of botanical and non-botanical adulterants such as pharmaceutical drugs, microbial contaminants, and inorganic materials. Advancement in microscopic techniques like attachment of digital video eyepiece with microscope and improvements in light, fluorescence, phase contrast, and scanning electron microscopes have improved the accuracy and capabilities of microscopy as a means of botanical authentication. Organoleptic analysis, used in combination with advanced microscopic equipment and attachment with computer, which is computer aided microscopy, provides further accuracy for botanical authentication and characterization of powder vegetable drugs.

Therapeutic Activity

Therapeutic activity of herbs is because of various constituents present in them. Therapeutic efficacy of medicinal plants depends upon the quality and quantity of chemical constituents which may vary depending on various factors, one amongst is the geographical localities which show quantitative variation in their chemical constituents. In some plants toxic constituents are also present therefore it is essential to evaluate their quality, safety and efficacy. Correct identification and quality assurance of the starting material is therefore an essential prerequisite to ensure reproducible quality of herbal medicine, which contributes to its safety and efficacy. In most of the cases of herbal medicine, misuse starts with wrong identification. Many of the traditional systems have records where one common vernacular name is given to two or more entirely different species.

Problems on Hand

Standardization of plant based medicine is a difficult task, because plants synthesizes not only single compounds but it may vary even up to hundreds of compounds may be present in plant. Hence it is difficult to standardize herbal medicines as compared to other medicines. Among consumers, there is a widespread misconception that "natural" always means "safe", and a common belief that remedies from natural origin are harmless and carry no risk. However, some medicinal plants are inherently toxic.

Further, as with all medicines, herbal medicines are expected to have side effects, which may be of an adverse nature. Some adverse events reported in association with herbal products are attributable to problems of quality. Major causes of such events are adulteration of herbal products with undeclared other medicines and potent pharmaceutical substances, such as cortico steroids and non-steroidal anti-inflammatory agents. Adverse events may also arise from the mistaken use of the wrong species of medicinal plants, incorrect dosing, and errors in the use of herbal medicines both by health-care providers and consumers, interactions with other medicines, and use of products contaminated with potentially hazardous substances, such as toxic metals, pathogenic microorganisms and agro-chemical residues.

The following examples demonstrate the range of problems encountered with the use of herbal medicines and products.

- Some herbal products were found to contain 0.1 - 0.3 mg of beta methasone per capsule after some patients developed cortico steroid-like side effects.
- Owing to misidentification of the medicinal plant species, plant materials containing aristolochic acid were used for manufacturing herbal products, which caused severe kidney failure in patients in several countries.
- Reports have been received by drug safety monitoring agencies of prolonged prothrombin times, increased coagulation time, subcutaneous hematomas and intracranial hemorrhage associated with the use of *Ginkgo biloba*.
- One of the most well-known traditionally used herbal medicines caused severe, sometimes fatal cases of interstitial pneumonia when used in conjunction with interferon.

Major Challenge

That must be overcome before herbs can join mainstream medicine is the quality of the literature in the field. Books, pamphlets, journals, and especially these days the Internet are filled with misinformation, much of it written to sell product, some of it written to express a point of view based on hope, not fact, or on misinformation. Most sites merely list herbs and their uses few mention regulation, safety, or efficacy. Even an herb with well-recognized toxicities, such as ephedra may have no cautionary statement. Another problem is that clinicians working with herbal products are still relatively unfamiliar with them often do not realize the necessity of adequate dosage from definition in the published papers. Many erroneous and unreproducible results have appeared in the medical literature because the clinicians accept at face value the quality of an herb that was adulterated, misidentified. In addition, they often fail to identify specifically, that is by scientific name, the botanicals in the product tested, as well as the precise dosage administered. Adverse events thus far reported in relation to herbal products are frequently attributable either to poor quality or to improper use, and it is therefore difficult to distinguish genuine adverse reactions to herbal medicines and herbal products until the cause of such events has been identified.

Scope of Researchers

Global estimate indicates that 80% of about four billion population cannot afford the products of western Pharmaceutical companies and have to rely on the use of traditional medicines which are mainly derived from plant materials. In spite of overwhelming influence and dependence on modern medicine and tremendous advances in synthetic drugs a large segment of world population still like drugs from plants. In many of the developing countries the use of plants drugs is increasing because modern life – saving drugs are beyond the reach of three quarters of the third world population, although many such countries spend 40–50% of their total health budget on drugs.

Now a days the interest of the manufacturing of herbal drug preparations is increasing because of many reasons such as:

1. Effectiveness of the drug or plant.
2. Easy availability.
3. Low cost.
4. Comparatively devoid of serious toxic effects.

Due to the background of all Vedas to Indian traditions, India unquestionably occupies the topmost position in the use of herbal drugs and is one of the foremost countries exporting the plant drugs and their derivatives.

Need for documentation of research work carried out on traditional medicines. These studies help in identification and authentication of the plant material. Correct identification and quality assurance of the starting materials is an essential prerequisite to ensure reproducible quality of herbal medicine which will contribute to its safety and efficacy. Simple pharmacognostic techniques used in standardization of plant material include its morphological, anatomical and biochemical characteristics. These standards are of vital importance not only in finding out gentry, but also in detection of adulterants in marketed drug.

Need for Clinical Trials

To gain public trust and to bring herbal product into mainstream of today health care system, the researchers, the manufacturers and the regulatory agencies must apply rigorous scientific methodologies and clinical trials to ensure the quality and lot-to-lot consistency of the traditional herbal products. Since the identities of the final products are not well defined and there are essentially no purification steps involved in the productions of herbal products, the quality and lot to lot consistency of the products rely mostly on the quality control of source materials and their manufacturing into the final products. Using modern technologies the quality and consistency of the heterogeneous herbal products can be monitored. A well-designed clinical trial is the method of choice to prove the safety and effectiveness of a therapeutical product. Manufacturers of the herbal products must adhere to the requirements of good manufacturing practices (GMPs) and preclinical testing before these products can be tested on human. The basic principle and design of the clinical trials for herbal products are the same as those for single component chemical product.

Widening the Herbal Options

It is a truism that going back to nature is something that has caught on world over, in the last decade. This has seen an upsurge in Ayurveda (herbal) products, an area where India's expertise dates back centuries. But it is only in the last decade that the country has truly seen the commercialization on the herbal concept. Herbal has now become a full-fledged wave, encompassing both beauty-care and health-care products. And herbal over the counter (OTC) drugs have gained substantial ground. Currently, according to industry estimate, of total pharmaceutical market of around Rs. 5, 000 crores, the total herbal market share is Rs. 700 crores, of which the OTC market constitutes around Rs. 400 crores.

The reason for the takeoff of herbal products has been gradual and can be ascribed to many reasons. The origins, according to many, can be sourced at the World Health Organization's Canberra conference in 1976, which promoted the concept of "traditional" medicines for the developed countries.

In India, almost a decade later, herbal cure began to be accepted, as alternative therapies with minimal side effect, gained in ascendance elsewhere.

The herbal drugs market itself is growing at a rate of between 20 and 30% annually, with individual companies registering different growth rates. The healthy growth rate of this marked can also be attributed to the Government's policy of encouraging the manufacturers of purely herbal products. This coupled with absence of any pricing guidelines, unlike "Drug Price Control Order (DPCO)" pricing guidelines for ethical drugs, has resulted in this segment being perceived as a highly lucrative alternative source of revenue. The new patent policy under "GATT" which will come into effect by the year 2005, has encouraged the herbal market. While the domestic market is opening up to the herbal phenomenon, the export market is also showing promise. Many pharmaceutical companies are targeting export as the prime source in the coming years. Today, the export market stands at Rs. 50 crore and is expected to expand to Rs. 20,000 crore in the next decade.

2

MEDICINAL PLANTS FOR ANTI-INFLAMMATORY

INTRODUCTION

Nature has provided a complete storehouse of remedies to cure all ailments of mankind. Since the down of civilization, in addition to food crops, man cultivated herbs for his medicinal needs. The knowledge of drugs has accumulated over thousands of years as a result of man's inquisitive nature, so that today we possess many effective means of ensuring health care. The history of herbal medicines is as old as human civilization. The documents, many of which are of great antiquity, revealed that plants were used medicinally in China, India, Egypt and Greece long before the beginning of the Christian era.

In Indian system of medicine, a large number of drugs of either herbal or mineral origin have been advocated for various types of diseases and other different unwanted conditions in humans. Ayurveda is one of the traditional systems of medicine practiced in India and Sri Lanka and can be traced back to 6000 BC. Ayurvedic medicines are largely based upon herbal and herbo-mineral preparations and have specific diagnostic and therapeutic principles.

Herbal medicines are a valuable and precious gift of the nature and have been playing a significant role in the prevention and treatment of various human ailments since the time immemorial. The major population of the south eastern Asian countries relies heavily on the efficacy of herbal remedies.

Plant-derived medicines are useful therapeutic options and often a safe form of therapy and in many instances specific phytomedicines have been shown to be clinically effective. There are many reasons for increased use of herbal medicines. These may range from appeal of product from 'nature' and the perception that such product are 'safe' (or at least 'safer than conventional medicines, which are often derogatorily referred to as 'drugs'), to more complex reasons related to philosophical views and religious beliefs of individuals.

Medicinal plants are accessible, affordable and appropriate source of primary healthcare for almost two-third population of our country. People who can't afford high cost of modern medicines are mainly dependent on these culturally familiar, simple, affordable and effective traditional medicines. Because of these favorable factors, there is wide spread interest in promoting traditional health system to meet primary healthcare needs. The spreading cost of modern medicines has made it necessary for the protection and promotion of the cultural and spiritual values of traditional medicines.

Inflammation

Inflammation is defined as the local response of living mammalian tissues to injury due to any agent. It is body defense reaction in order to eliminate or limit the spread of injurious agent as well as remove the consequent necrosed cells and tissues.

Inflammation is the reaction of vascularized living tissue to local injury. Inflammation is the name given to the more-or-less stereotyped ways our tissues respond to noxious stimuli, with blood vessels and white blood cells as its twin centerpieces and a host of proteins as actors. Inflammation "destroys, dilutes, or walls off the injurious agent" and sets in motion the limited powers of the body to heal itself.

Depending upon the defense capacity of the host and duration of response, inflammation can be classified as acute and chronic inflammation.

Causes of Inflammation

Various agents may kill or damage cells as

1. Physical agents like Heat or cold, Trauma, Radiation.
2. Chemical agents like Simple chemical poisons and Organic poisons.
3. Infective agents like Bacteria, Virus, Parasites etc.
4. Immunological agents like Antigen – Antibody, Cell mediated.

Treatment of Inflammation

In the treatment of inflammation generally non-steroidal anti-inflammatory drugs (NSAID) are used. The drugs of choice are as follows.

1. Non-steroidal anti-inflammatory drugs (NSAID): Aspirin, Paracetamol, indomethacin, Sulindac, diclofenac and mefenamic acid.
2. Anti-inflammatory drugs without direct analgesic action: Glucocorticoids like prednisolone.
3. Remission inducing drugs (disease modifying Antirheumatic drugs: Chloroquine, penicillamine, sulphasalazine.
4. Immuno suppressant drugs: Methotrexate, azathioprine, cyclosporine.

In the recent era, in spite of these drugs herbal drugs are becoming more popular for the treatment of inflammation which have minimum adverse effects as compared to above mentioned drugs.

Common Adverse Effects of NSAIDs

Platelet Dysfunction, Gastritis and peptic ulceration with bleeding, Acute renal failure in susceptible individuals, Sodium+ water retention and edema, Analgesic nephropathy, Prolongation of gestation and inhibition of labor during delivery, Hypersensitivity (not immunologic but due to PG inhibition), Cardiovascular thrombosis.

Mechanism of Action of NSAIDs

NSAIDs are said to be polyvalent in action, because they are able to modulate more than one molecular or cellular event, which is associated with inflammatory event. NSAIDs have many biochemical actions. They are enumerated as follows:

1. Inhibition of formation of inflammatory mediators' *i.e.* histamine, serotonin, prostaglandin, kinins.
2. Moderation of activity of inflammatory proteases.

3. Inhibition of oxidative phosphorylation, depriving the inflamed tissue of needed metabolic energy in the form of adenosine triphosphate (ATP).
4. Displacement of anti-inflammatory peptide from albumins.
5. Hyper polarization of neuronal membrane in the acidic environment of inflamed tissue.
6. Stabilization of lysosomal membrane.

Uses of NSAIDs

(i) Anti-inflammatory.
(ii) Analgesic.
(iii) Antipyretic.
(iv) Treatment of gout.
(v) Prophylaxis of heart disease (myocardial infarction and stroke).
(vi) Prophylaxis of colorectal cancer.
(vii) Treatment of Alzheimer's disease.

A. Acute Transient Phase

1. Changes in vascular caliber and blood flow.
2. Increased vascular permeability resulting in the formation of inflammatory exudates and local edema.

B. Delayed Sub-Acute Phase

Migration of leucocytes and phagocytes from blood into the extra vascular tissues.

C. Chronic Proliferative Phase

Tissue degeneration and fibrosis.

Inflammation can also be described on the basis of the cardinal signs.

Calor

Due to increase of blood flow at the site of tissue injury, their is an increase of the local temperature.

Rubor

Involves dilation of small blood vessels, arterioles, capillaries and venules within the injured area.

Tumor

The dilatation of blood vessels result in increased permeability of the vessels walls, there by allowing protein rich fluid (containing high amount of leucocytes) to escape into the tissues of the damaged area - leading to local oedema formation.

Dolor

Inflammation results in release of endogenous mediators, particularly histamine, serotonin, bradykinin, prostaglandin (and arachidonic acid metabolites). Such mediators elicit pain response even when released in minute quantities. Pain may also result from the increased tissue tension caused by inflammatory exudates (oedema formation). Thus evacuation of pus from wounds, abscesses and boils may result in immediate lowering of pain.

Loss of function

Pain may result in diminished muscular movement and such limitation of muscular movement is aggravated following oedema or swelling in the inflamed area.

In order to understand the inflammatory process, it is essential to have a broad idea on the endogenous mediators which play a significant role in this inflammatory process.

Histamines and serotonin (5 hydroxytryptamine 5HT) play a predominant role in acute inflammation. Thus histamine antagonist hale emerged as effective drugs in controlling urticarias and asthma.

Kinins are a group of peptides released in blood by kallikreins. The kinins generated locally, contribute in the acute and possibly chronic phase of inflammation thus producing vasodilatation, local oedema and pain. Kinins also mediate leukocyte migration.

Finally, prostaglandins (products of oxidative metabolism of arachidonic acid) play a major role in the entire inflammatory process. The arachidonic acid is metabolized by two pathways - cyclooxygenase (produce prostaglandins, thromboxanes - elicit inflammatory response) and lipooxygenase pathway (produces leukotrienes; leukotriene B4 in particular is a highly potent inflammatory mediator). These products are hyperanlgesic, potent vasodilators and also contribute in eryrneme, oedema and pain. The currently available nonsteroidal anti-inflammatory drugs (NSAID's) are potential inhibitors of this cyclooxygenase pathway of arachidonic acid metabolism.

Platelet activating factor (Paf-acether) is one of the recently reported mediator believed to be involved significantly in the inflammatory process. The mediator is a phospholipid, formed by different cells, such as eosinophils, macrophages, platelets and neutrophils. It is synthesized from membrane phospholipids with the help of phospholipase A2. Paf is known to stimulate free arachidonic acid from cells and also metabolize it to eicosanoids.

Finally, the involvements of free radicals have been recently marked as an emerging area in understanding the inflammatory process. Free radicals (H_2O_2, dl-i, HOCI) can be described as chemical species, with one of more unpaired electrons in the outer most orbit. These chemical substances are essential to our biological system for counteracting the microbes entering into our system, in millions but these are destructive unless tightly controlled. There are a number of scientific reports related to the trials with free radical scavenger (enzymes such as superoxide dismutase, catalase: iron chelators Vit.E; plant products); such studies show encouraging results with the scavengers described above in suppressing both acute and chronic inflammation.

Chronic inflammatory disorders are classified on the basis of changes in connective tissues. These include rheumatic fevers, rheumatoid arthritis, ankylosing spondylitis, polyarteritis nodosa, systemic lupus erythematosus, and osteoarthritis. In rheumatic fever, the exudative changes characteristic of acute inflammatory condition is accompanied with proliferative changes characteristic of chronic cammation. The predominant feature of chronic inflammatory concentration characterized by proliferation of fibrous tissue, though tie euthtive changes (as observed in acute inflammation) are also evidenced and it is often observed that chronic inflammation is preceded by acute changes.

Rheumatoid arthritis is an autoimmune disease where there is a complex interaction of genetic, immunological and local factors in the initiation of the disease. It is a relatively common disease which affects the joints; it is a combination of chronic inflammation fibrosis and tissue degeneration. Studies have revealed that prostaglandin E2 is associated with the analgesic response in rheumatoid arthritis. Recent studies also implicate leukotrienes and PAF in tissue degeneration and fibrosis associated with rheumatoid arthritis. Studies also indicate involvement of monokines including interleukin 1 and tumor necrosis factor (TNF) in synovial fibrosis and tissue damage, associated with the chronic proliferative phase of inflammation. There is a wide scope for development of suitable monkine synthesis inhibitors or receptor antagonists which in turn may be useful for treatment of this dreaded disorder.

The salicylates are one of the earliest known drugs used in the treatment of inflammatory disorders. Hippocrates, about 2400 years ago recommended the use of willow bark for the treatment of eye disease and pain during childbirth. In the Papyrus Ebers (1550 BC) it is stated that a remedy to expel rheumatic pains in the womb is to apply the dried leaves of myrtle (contain salicylates) prepared with beer, to the sacral and hypo gastric regions. In China, preparations from the bark of *Populus alba* and decoctions from the young shoots of *Salix babylonica* have been reported to be used since several hundred years in the treatment of rheumatic fevers, colds, hemorrhages and goiter. The drugs used in inflammatory disorders may be classified as

1. Drugs with analgesic but with negligible anti-inflammatory actions Aniline derivative paracetamol is the principal member of this group now used; phenacetin is obsolescent.
2. Drugs with analgesic and mild to moderate anti-inflammatory actions:
   (a) Propionic acid derivatives which include ibuprofen, ketoprofen, fenoprofen, naproxen, flurbiprofen.
   (b) Anthranilic acid derivatives which include mefenamic acid, flufenamic acid.
   (c) Acryacetic acid derivatives which include fenclofenac, diclofenac.
3. Drugs with analgesic and marked anti-inflammatory actions -
   (a) Salicylates and derivatives which include aspirin, sodium salicylate, benorylate, diflunisal, aloxiprin, salsalate.
   (b) Pyrazolone derivatives which include phenylbutazone, oxyphenbutazone, azapropazone, feprazone.
   (c) Indole derivatives which include indomethacin, sulindac.

The above is not a rigid classification since effect also depends upon the dose used, but for the groups of drugs it generally holds true.

As we often observe, inflammatory conditions require prolonged treatment with anti-inflammatory drugs. The currently used non steroidal anti-inflammatory drugs have several toxic side effects which limit their use in long term therapy.

These drugs share several unwanted effects. The most common is a propensity to induce gastric or intestinal ulceration that can sometimes be accompanied by a secondary anemia from resultant blood loss. These drugs vary considerably in their tendency to cause such erosions. Gastric damage by these agents can be brought about by at least two distinct mechanisms. While local irritation by the drugs inthe stomach allows back diffusion of acid into mucosa and induce tissue damage.

The predominant prostaglandins synthesized by the gastric mucosa are PGE_2 and PGI_2 these eicosanoids inhibit acid secretion by the stomach and promote the secretion of cytoprotective mucus in the intestine. Such prostaglandins and their analogues can prevent mucosal damage including that induced by anti-inflammatory drugs. Thus, inhibition of the synthesis of endogenous prostaglandins may render the stomach more susceptible to damage.

Inflammation - Inflammatory disease including different types of rheumatic disorders is a major concern to scientists around the globe. Today a considerable percentage of the human population is affected by this disease. A variety of anti-inflammatory drugs are flooding the world market today but a very few are relatively non-toxic and fit for long term consumption. Moreover, discontinuation of drug therapy in chronic inflammatory conditions often leads to reappearance of symptoms.

A gastrointestinal disorder, related to the use of anti-inflammatory drugs is a perennial problem that irks medical world even today. Modern science are seen the advent of many new anti-inflammatory drugs. But also! the problem still remains.

Thus, it can be mentioned that intensive research with indigenous drugs can definitely open up new vistas in inflammation therapy and purified natural compounds may serve as template for synthesis of a new generation of anti-inflammatory drugs which in turn may have low toxicity and better therapeutic index.

PLANTS FOR ANTI-INFLAMMATORY

1. *Acanthus ilicifolius* Linn. (Acanthaceae)

Common names

Sans.: Harikusa
Beng.: Hargoza
Hindi: Harkuch
(Hargoza) Kanta.
Bomb: Nivagur

Distribution

A gregarious, sparingly branched, evergreen shrub 0.6-1 5 m in height, common in the tidal swamps of creeks and rivers along the East and West coasts: also distributed in Meghalaya and the Andamans.

Parts used: Leaves.

Pharmacological Activities

Leaves are used as fomentation in neuralgia and rheumatism.

Chemical Constituents

Analysis of fresh leaves gave the following values (dry basis):

Water 78.20, Carbohydrates 20.95, lipids 13.55, protein nitrogen 1.87, non-protein nitrogen 0.24 and ash 12.2%, Cobalt 3.3, Copper 9.0, Manganese 96.7 Molybdenum, 16.7 and Zinc 24.7 mg/g. A new alkaloid, acanthicifolin has been isolated from the air-dried plant. The plant also contains flavones.

2. *Alangium salviifolium* (L.F.) Wang. (Alangiaceae)

Common names

Eng.: Sage leaved Alanglum
Hindi: Dera

Distribution

A small tree with edible fruits. Found throughout the drier parts of India, especially in forest of South India.

Parts used: Stem bark.

Pharmacological Activities

Anti-inflammatory activity of the bark extract was found in formaldehyde induced arthritis and Granuloma pouch oedema in animals.

Chemical Constituents

A new alkaloid dimethylcephaeline along with cephaeline. Psychatrine tubulosine and dimethylpsychotrine isolated from stem bark: substance AL 60. previously reported to have hypotensive activity, shown to be a mixture of psychorine, cephaeline and demethylcephaeline; isolation and characterisation of new steroi stigmast-5,22,25-trien-3b(Beta)-oil from leaves; a mono terpenoid lactam-alangiside isolated and its structure elucidated; N-benzoyl-L-phenylalanine, m.p. 169, was isolated: structures of new D. E-cis fused neohopane derivative As- alangidiol and Isoalangidiol; stereostructure and synthesis of (±) alangicine: structure and relative configuration of (±) alangimarckine determined by its partial synthesis.

3. *Allium sativum* Linn. (Liliaceae)

Common names
Eng.: Garlic
Hindi: Lasun
Beng.: Rasun

Distributions

A hardy perennial, 60 cm in heightfit, native to central Asia and cultivated all height fit India

Parts used: Bulbs.

Pharmacological activities

Garlic is an effective long term preventive treatment for all rheumatic and catarrall conditions. It produces anti-inflammatory activity against formalin induced arthritic in albino rats. A concentrate containing the active principle, allicin and allinase proved effective in the treatment of rheumatoid arthritis.

Chemical constituents

Isolation of biologically active compound scordinin Al which on alkaline hydrolysis yielded a peptide, scormine and allylthiofructosiduronic acid; fine unidentified saponins found; garlic bulbs yielded a mixture of polysaccharides containing pectic acid, a D-galactone and a fructan component which contained fructose (94.4) and glucose (4.3%); a linear inclined type structure suggested for fructan on basis of methanolysis. Garlic also contains allicin and alliinase.

4. *Alpinia calcarata* Rosc. (Zingiberaceae)

Common names
Eng.: Shell ginger
Mal.: Kattuchona, Oriya Toroni
Tam.: Amkolingi

Distribution

It is a slender, rhizomatous herb, often cultivated in gardens in eastern and southern India for its white flowers, variegated with red and yellow in pyramidal panicles.

It is frequently available in the sub-Himalayan region of Bihar, West Bengal and Assam, and is extensively cultivated all over India mostly in shady situations.

Parts used: Leaves, flowers.

Pharmacological activities

The herb is reported to possess anti-tubercular properties. The water-soluble fraction of the alcoholic extract of the air-dried plant in reported to exhibit a significance anti-inflammatory activity in albino rats similar to the betamethasone.

Chemical constituents

The leaves yield an essential oil containing mostly methyl cinnamate.

5. *Anacardium occidentale* L. (Anacardiaceae)

Common names
Eng. Cashew nut
Beng.: Kaju Badam
Hindi: Kaju

Distribution

A small evergreen tree, native of tropical America from Mexico to Peru and Brazil but now cultivated largely in Malabar, Kerala, Karnataka, Tamil Nadu and Andhra Pradesh and to some extent in Maharashtra, Goa, Orissa and W. Bengal.

Parts used: Whole plant.

Pharmacological activities

Anti-inflammatory activity of the plant extract was found in carrageenan induced oedema, Cotton pellet induced granuloma, formaldehyde induced arthritis and Freund's complete adjuvant induced arthritis in animals.

Chemical constituents

See Antidiabetic chapter.

6. *Anemone obtusiloba* D. Don. (Ranunculaceae)

Common names
Eng.: Al-Kanet, Kumaon-Kakriya, Ratanjota
Punjabi: Padar, Rattanjog

Distribution

Temperate and alpine Himalayas from Kashmir to Sikkim between altitudes of 2,100 and 4.200 m, and in the Nilgiri hills, above an altitude of 1,800 m.

Parts used: Seed.

Pharmacological activities

Oil extracted from the seed is used in rheumatism.

Chemical constituents

The air-dried plant is reported to contain a substance similar to anemonin.

7. *Anthemis cotula* Linn. (Asteraceae)

Common names
Eng.: Cotula, Dog Fennel, Mayweed, wild Chamomile.

Distribution

Naturalized in waste places in Uttar Pradesh and Himachal Pradesh.

Parts used: Aerial part of the plant.

Pharmacological activities

The aerial part of plant provides relief in inflammation of tissues.

Chemical constituents

The whole fresh plant yields 0.01% of reddish, bitter essential oil which is acidic in reaction. A crystalline acid (imp. 58°) is present in both free and ester form in the volatile oil. The flowers and leaves contain alkaloids. An optically active sesquiterpene lactone isolated.

8. *Arnebia hispidissima* Forsk. (Boraginaceae)

Distribution

A diffuse or prostrate, xerophytic herb distributed(North West India) in upper gangetic plains, Punjab, Rajasthan and Gujarat.

Parts used: Whole plant.

Pharmacological activities

Plant extract showed anti-inflammatory effects in carrageenan induced oedema, cotton pellet induced granuloma in animals.

Chemical constituents

Flavonoids (Vitexin) were isolated. Roots yielded di-alkanin (Shikalkin) as a crystalline red solid.

9. *Azadirachta indica* A. Juss. (Meliaceae)

Common names
Eng.: Margosa tree
Beng.: Nim
Hindi: Nim, Neem

Distribution

A common tree in the plain of West Bengal and other regions in the plains of India.

Parts used: Bark.

Pharmacological activities

Sodium nimbinate isolated from plant possessed potent anti-inflammatory activity in carrageenan induced oedema and formaldehyde induced arthritis in animals.

Chemical constituents

From the plant oil three bitter principle was isolated 1) Nimbin a sulpher free neutral, water insoluble colorless crystalline product (m.p. 205°, yield 0.1%), 2) Nimbidin (m.p. 192°, yield 0.01%), 3) Nimbidin (m.p. 90-100°, yield 1.1%) a cream coloured powder insoluble in water. A new oxophenol-nimbiol, m.p. 250° in addition to known substance nimbosterol, m.p. 82° from trunk bark was isolated.

10. *Berberis asiatica* Roxb. (Berberidaceae)

Common names
Beng.: Daruharidra, Garhwal Kuigora
Sans.: Daruharidra

Distribution

Is a pretty shrub 1.8 to 2.4 m in height, armed with trifid spines. Commonly occurring in the Himalayas from Himachal Pradesh at 600-2.700 m eastwards to Bhutan and Assam at 1,500-1.800 m, and on Pareshnath hills in Bihar, Pachmarhi in Madhya Pradesh and Mount Abu in Rajasthan. It also grows in hedges.

Parts used: Stems.

Pharmacological activities

The root has bitter, sharp, hot taste, as a fomentation, removes inflammation and swelling.

Chemical constituents

The alkaloids are present in the plant. Berberine and palmatine are present as chlorides in addition, two more chemicals are present in small amount, in the root.

11. *Berberis petiolaris* Wall. (Berberidaceae)

Common name
Arabic: Amberbaris
Punjab: Chochar
Urdu: Amber

Distribution

Western Himalayas from Kashmirto Nepal upto 12,000 ft. Garhwal and Kumaun at 2400 m.

Parts used: Roots.

Pharmacological activities

The root has cooling effects, used in paralysis and rheumatism.

Chemical constituents

Berberine, berbericine and a new more polar alkaloid isolated as picrate, m.p. 160° from roots.

12. *Bergenia ligulata* Engl. (Saxifragaceae)

Common name
Sans.: Pashanabheda, Shilabheda.

Distribution

Distributed in South and Eastern Asia.

Parts used: Roots (Rhizomes).

Pharmacological activities

Acetone extract of the rhizomes possesses potent anti-inflammatory activity but the activity decreases with increasing dosage.

Chemical constituents

(-)Afzelechin isolated from roots. Saxin isolated from roots was identified as bergenin. Bergenin, its C-glycoside, Beta-sitosterol and (+) Catechin-3-gallate isolated from roots.

13. *Boerhaavia diffusa* L. (Nyctaginaceae)

Common names
Eng.: Horse-purslane, Hog weed
Beng.: Punarnava
Hindi: Punarnava

Distribution

A herb distributed throughout India.

Parts used: Whole plant.

Pharmacological activities

Plant possesses good anti-inflammatory activity.

Chemical constituents

Hentriacontane, Beta-sitosterol and ursolic acid isolated from roots. A glycoprotein with a molecular weight of 16000-20000 daltons isolated from roots. Ash (11.8%), calcium (1.2%), and potassium (2.3%), presence of alkaloids free and combined amino acids determined in aerial parts of plant.

14. *Boswellia serrata* Roxb. (Burseraceae)

Common names
Eng.: Indian Oil banum Tree
Hindi and Beng.: Salai

Distribution

A medium to large sized, deciduous balsamiferous tree up to 18 m height and 2.4 m in girth commonly found in dry forests from Punjab to West Bengal and in Peninsular India. The tree is common at the foot of western Himalayas, in Rajasthan, Gujarat, Maharashtra, Madhya Pradesh, Bihar, Orissa, Andhra Pradesh and further south in Peninsula.

Parts used: Barks

Pharmacological activities

The deflated extract of gum exudates (oleo-gum-resin) was found to possess marked anti-inflammatory and anti-arthritic activity in animals (unpublished report).

Chemical constituents

Oleo-gum-resin consists of three principal constituents namely turpentine liquid, rosin like resin and gum. Oil resemble-sturpentine oil. 3c-Hydroxytirucall-8,24-dien-21-oic acid its acetyl derivative, 3-Ketotirucall-8,24-dien-21-oic acid and 33- hydroxytirucall-8,24-dien-21-oic acid isolated form resin.

Fresh leaves on steam distillation gave an essential oil. Non-volatile fraction (gum-rosin) of oleo-gum-resin yielded a new diterpenic alcohol.

15. *Bryophyllum pinnatum* Lam. (Crassulaceae)

Common names
Beng.: Koppata
Hindi: Zakhm-haiyat
Guj.: Ghayamari.

Distribution

Bryophyllum pinnatum is believed to be a native of tropical Africa, naturalized throughout the tropics of the world. It is cultivated in garden and also grows profusely on the hills of North Western India, Deccan and Bengal.

Parts used: Leaves.

Pharmacological activities

Anti-inflammatory activity of the plant extract was found in carrageenan induced oedema cotton pellet granuloma, formaldehyde induced arthritis, freund's adjuvant induced oedema, Turpentine induced arthritis in animals.

Chemical constituents

Quercetin-3-L-rhamnoside-L-arabinofuranoside isolated. Quercetin-diarabinoside, m.p. 1900 and kaempferol-3-glucoside, waxes, flavonoid glycoside and rhenolic compound isolated. Alkanes C_{25}-C_{35}, alkanols C_{26}-C_{34}, α-amyrin, Beta- amyrin and sitosterol isolated from non-saponihydroxybenzoic acids, quercetin and kaempferol detected in leaves; wax hydrocarbons (C_{25}-C_{35}) wax alcoholic (C_{26}-C_{34})and fatty acids obtained from wax of leaves.

16. *Calophyllum inophyllum* L. (Clusiaceae)

Common names
Eng.: Alexandrian laurel
Beng.: Sultana champa
Hindi: Surpun, Sultan champa

Distribution

An evergreen tree distributed on the sea shores of India, particularly Orissa, Karnataka, Maharashtra and the Andamans also cultivated as an ornamental tree.

Parts used: Seeds.

Pharmacological activities

Seeds yield a fixed oil known as Domba, applied externally in rheumatism. Seed extract showed anti-inflammatory activity in granuloma pouch, carrageenan induced oedema, cotton pellet induced granuloma, formaldehyde induced arthritis, Freund's complete adjuvant induced arthritis in mice and rats.

Chemical constituents:

Callophylloeide, Xanthones (Dehydro-cycloguanadine, Callophyllin-B, Sacareubin, 6-deoxyjacareulin) isolated. 4-Phenylecoumarins-calophyllolide, inophyllolide and calophymic acid from ripe seeds; cinnamic acid, inophyllic and calophyllic acids and a new 4-phenylcoumarin-ponnalide from unripe seeds; a new myricetin glucoside, myricetin and quercetin from androecium; leucocyanidin from petals; friedelin, 3 new triterpenes-canophyllal, canophyllol and canophyllic acid (+) inophyllolide, m.p 188°, its cis isomer, m.p. 149° and 12-hydroxy derivative of cis isomer, m.p. 200°, from leaves; inophylloidic acid from bark resin; jacareubin, 6-deoxyjacareubin and 2-(3,3-dimethylallyl)-1,3,5,6-tetrahydroxyxanthone from heartwood.

17. *Cananga odorata* Lam. (Annonaceae)

Common names
Eng.: Ylang-ylang
Tamil: Karumugai
Tel.: Chettusampangi
Malaya.: Kananga

Parts used: Flowers.

Pharmacological activities

Flowers of the oil are used for application of Cephalalgia,z1ophthalmio and gout.

Chemical constituents

The flowers yield "ilang-ilang" of perfumes "Cananga oil", consists of the early portions of the distillate. Canangine was isolated and found identical with eupolauridine.

18. *Canscora decussata* Schult. (Gentianaceae)

Common names
Beng.: Dankuni
Hindi: Sankhaphuli

Distribution

A herb available throughout India.

Parts used: Whole plant.

Pharmacological activities

Fresh juice of the plant is recommended in epilepsy, insanity and nervous debility. Extract of the plant showed anti-inflammatory activity in carrageenan induced oedema, cotton pellet induced granuloma, formaldehyde induced arthritis in animals.

Chemical constituents

See anti-fertility chapter.

19. *Cassia alata* Linn. (Caesalpiniaceae)

Common names
Eng.: Ringworm Senna
Beng.: Dadmardan, Dadmari
Hindi: Dadmardan, Dadkapat, Benanakhi, Vilaytie aghatea

Distribution

A large hand some shrub or a small tree, 1-5 m in height introduced from the West Indies and cultivated in the gardens, and also found wild almost throughout India and in the Andaman Islands.

Parts used: Whole plant.

Pharmacological activities

The plant is reported to possess anti-inflammatory properties.

Chemical constituents

The leaves contain cassiaxanthone, kaempferol, and its glycoside, aloe-emodin, chrysophanol, rhein, physicion-1-glucoside and 13-sitosterol. The roots contain quinine derivatives. Crysophenol, emodine, rhein and aloe-emodine isolated from leaves and fruits; glycoside of rhein, aloe-emodine and emodine obtained from leaves and roots.

20. *Cassia fistula* Linn. (Caesalpiniaceae)

Common names
Eng.: Golden-shower, Indian Laburnum
Beng.: Amaltas
Hindi: Bendralathi

Distribution

A deciduous, medium sized tree up to 24 m in height and 1.8 m in girth, cultivated almost throughout India. The tree is one of the most wide spread in the forests in India, usually occurring in deciduous forests throughout the greater part of India, ascending up to an altitude of 1,220 m in the sub-Himalayan tract and outer Himalayas. It is common throughout the Gangetic valley, particularly abundant in the bhabar tracts. Central India and South India.

Parts used: Roots, barks.

Pharmacological activities

The aqueous extract of the root bark exhibits anti-inflammatory activity.

Chemical constituents

The root bark yields a mixture of three flavonoids, one of them was identified as fistucacidin.

The root bark also contains tannins, phlobaphenes, reducing sugars, and oxyanthraquinones. Fruits pulp contain rhein, glucose, sucrose, fructose, and pod contain fistulic acid and leucopeler gonidin. Fistucacidin m.p. 245 obtained from bark.

21. *Cedrus deodara* Roxb. (Pinaceae)

Common names
Eng.: Himalayan cedar, Deodar
Beng.: Devadaru
Hindi: Deodar.

Distribution

A tall, evergreen tree distributed in N.W. Himalayas from Kashmir to Garhwal. Forests of deodar occur in Kulu, Kashmir, Chamba, Tehri-Garhwal, Almora, Simla, Chakrata and Mussoorie hill stations.

Parts used: Bark, leaf.

Pharmacological activities

The bark is useful in fever and rheumatism. All parts bitter, hot, pungent, light, oleagenous, useful in belching inflammations. The leaves lessen inflammation.

Chemical constituents

Flavonoids ascorbic acid, etheral oil (0.056%) isolated from plant. Wood yields an oleoresin and darkcolored oil. Isolation and characterization of α-himachalene, bp. 93°12 mm. and Beta-himachalene, b. p. 121 /4 mm. from essential oil (+) lengiborneol and two new sesquiterpene alcohols himachalol, m.p. 67° and allohimachalol, m.p. 85° from essential oil; deodarin, m.p. 248° from the stem bark. Centdarol isolated and characterised as 23,7 3-dihydroxy-himachal-3-ene; isolation and structure of isocentdarol; structure determination of oxidohimachalene isolated from essential oil; structure of isohimachalane, isolated from wood essential oil; structures of deodarone and limone carboxylic acid isolated from wood.

22. *Chrysanthemum indicum* L. (Asteraceae)

Common names
Eng.: Japanese chrysanthemum
Beng.: Chrysanthemum
Hindi: Guldaudi
Mar.: Shevati

Distribution

Native to China and Japan cultivated in India as ornamental.

Parts used: Leaves.

Pharmacological activities

Leaves are useful in migraine.

Chemical constituents

dl-camphor, azulene and Beta-3-carene, B.P. 70°/30 mm, obtained by distillation of flowers; chrysanthenone from essential oil; isolation of a new sesquiterpene lactone-yehuja lactone along with chamazulene from flowers.

23. *Cimicifuga foetida* Linn. (Ranunculaceae)

Common names
Eng.: Bugbane
French: Acteefetide, cimicaire
German: Wanzen kraut
Punjab: Jiunti

Distribution

Temperate Himalayas from Kashmir to Bhutan (7,000-12,000 ft), E. Europe, Siberia.

Parts used: Roots.

Pharmacological activities

The root is poisonous. In Europe the root is considered a mild emetic, purgative. In China and Indo-China it is used as anti-periodic and prescribed in rheumatic affections.

24. *Cocculus hirsutus* L. (Menispermaceae)

Common names
Sans.: Vasanti tikta
Beng.: Huyer
Hindi: Jamti ki bel, Kharetaki-bel

Distribution

A climbing shrub occurring throughout tropical and sub-tropical tracts of India from the foot of Himalayas to South India.

Parts used: Roots, stems.

Pharmacological activities

Roots are useful in chronic rheumatism and venereal diseases. Extract of stems and roots are sedative spasmolytic. Roots are used in stomach ache in children.

Chemical constituents

D-Trilobine and DL-coclaurine from roots; 13-sitosterol, ginnol and monomethyl ether of inositol, m.p. 226° isolated; essential oil,

b. p. 127° and two alkaloids were also isolated but not characterized.

25. *Crotalaria laburnifolia* Linn. (Papilionaceae)

Common names
Hindi: Muna
Tel.: Pedda-galligista

Distribution

W. Bengal and some other places of India.

Parts used: Whole plant.

Pharmacological activities

Possessed potent anti-inflammatory activity in carrageenan induced oedema, cotton pellet induced granuloma, formaldehyde induced arthritis. It also possessed anti hyaluronidase activity.

Chemical constituents

An alkaloid-crotalaburnine, m.p. 185°, a pigment-lutexin and Beta-sitosterol isolated from seeds; apyrrolizidine alkaloid-anacrotine, m.p. 197° isolated from seeds, shown to be a cyclic diester of senecic acid.

26. *Cryptolepis buchanani* Roem. and Schult. (Asclepiadaceae)

Common names
Hindi: Karanta
Tel.: Adavipalatige

Distribution

Throughout India.

Pharmacological activities

Latex of the plant mixed with hot water is applied on knees to cure rheumatism.

Chemical constituents

Isolation and structure of a pyridine alkaloid-buchananine.

27. *Curcuma domestica* Val. (Zingiberaceae)

Common names
Eng.: Turmeric
Beng.: Halud
Hindi: Haldi

Distribution

A perennial herb cultivated mainly in Tamil Nadu, Andhra Pradesh, Maharashtra, Bihar, Kerala, Orissa.

Parts used: Rhizomes.

Pharmacological activities

Wide medicinal uses. It is used as stomachic toxic, blood purifier, antiseptic, also applied in sprains.

Chemical constituents

See Anti-fertility Chapter.

28. *Cyperus rotundus* L. (Cyperaceae)

Common names
Eng.: Nut grass
Beng.: Muthaghas
Hindi: Motha

Distribution

A perennial sedge distributed throughout India.

Parts used: Whole plant.

Pharmacological activities

Plant extract showed potent anti-inflammatory activity in carrageenan induced oedema, cotton-Pellet induced granuloma in rats.

Chemical constituents

Beta-Sitosterol and an essential oil. Cyperene-1 (a tricyclic sesquiterpene) and cyperene-2 (a bicyclic sesquiter pene hydrocarbon) isolated from tubers; patchoulenone, m.p. 52°; a new sesquiterpene ketone-mustakone, bp. 128°/i mm from essential oil, has same skeleton as copaene; isolation, structure and absolute configuration of cyperotundone; m.p. 46°, from tubers, cyperolone, m.p. 41°, from tubers; a new sesquiterpenoid-sugetriol triacetate, m.p. 132° from tubers of Japanese nutgrass; 27 compounds separated from essential oil by GLC and four of these copadiene, b. p. 130°f 1 mm, epoxyguaiene, bp. 102°/i mm, rotundone, bp. 128°/i mm, and cyperolone, b. p. 120°/0.1 mm - characterised; sesquiterpenic ketolsugeonol; structure and absolute configuration of cyperol and isocyperol; structure of α-rotunol, m.p. 87°, and 13-rotunol, m.p. 118°; two nor sesquiterpenoids - kobusone and isokobusone.

29. *Dalbergia lanceolaria* L.F. (Papilionaceae)

Common names
Beng.: Chakemdia
Hindi: Bithua
Mar.: Dandous

Distribution

A tall deciduous tree available throughout India.

Parts used: Whole plant.

Pharmacological activities

Plant extract showed potent anti-inflammatory activity in carrageenan induced oedema and formaldehyde induced arthritis in rats.

Chemical constituents

Lanceolarin, m.p. 165° from root bark, characterized as biochanin A-7-apiosyl-glucoside; psi-baptigenin isolated from flowers and leaves identified as 7-hydroxy-3',4'-methylene-dioxysoflavone.

30. *Desmodium gangeticum* D.C. (Papilionaceae)

Common names
Beng.: Salpani
Hindi: Sarivan
Sans.: Shalaparni

Distribution

A shrub distributed throughout India.

Parts used: Whole plant.

Pharmacological activities

Anti-inflammatory and antipyretic activity was shown in carrageenan induced oedema, cotton pellet induced granuloma, writhing response and pyrexia in rats by the plant extract.

Chemical constituents

A new pterocarpan-gangetin-isolated and characterized as 7α, 1 2α-dihydro-1 3-methoxy-3-3-dimethyl-1 I -(3-me- thyl-2-butenyl)-3H,7H-benzofuro[3,2-C]pyrano[3,2-g] benzopyran-10- 0l; detection of 5 phospholipids in seeds by TLC; twelve alkaloids of four structural types (carboxylated and decarboxylated tryptamine, 3-carbolines and Beta-phenethylamines) isolated; two pterocarpanoids - gangetinin and desmodin isolated and their structures determined.

31. *Dysoxylum binectariferum* Hook. F. (Meliaceae)

Common names
Beng.: Lassuni
Tam.: Agunivagil
Kan.: Agilu

Distribution

Sikkim, Assam, W. Bengal, Western Ghats and Andaman Island.

Parts used: Whole plant.

Pharmacological activities

Anti-inflammatory activity of the plant extract was tested in carrageenan induced oedema in animals.

Chemical constituents

A new tetranortriterpene - dysobinin - isolated from fruits and characterized.

32. *Echinops echinatus* Roxb. (Asteraceae)

Common names
Sans.: Kantalu
Hindi: Utakanta
Mar.: Kadechubak

Distribution

A herb occurring throughout India.

Parts used: Roots and other parts.

Pharmacological activities

Plant extract showed anti-inflammatory activity in acute carrageenan paw edema, formaldehyde induced arthritis and Adjuvant-induced acute and chronic arthritis in rats.

Chemical constituents

Isolation of Beta-amyrin and lupeol.

33. *Elephantopus scaber* L. (Asteraceae)

Common name
Beng.: Gojialata
Hindi: Gobhi
Sans.: Gojihva

Distribution

Throughout hotter parts of India.

Pharmacological activities

Plant paste without sugar is applied in rheumatism.

Chemical constituents

Epifriedelinol, lupeol, stigmasterol and a mixture of triacontan-1-ol and doltriacontan-1-ol isolated. A new ditactone - isodeoxyelephantopin isolated.

34. *Flacourtia indica* Merr. (Flacourtiaceae)

Common names
Beng.: Baichi
Hindi: Bailangra
Sans.: Svadukantaka

Distribution

Sub-Himalayan tract ascending up to 1200 m, Indusplains, upper gangetic plains.

Parts used: Seeds.

Pharmacological activities

Seeds are made into paste and used in rheumatism.

35. ***Garcinia mangostana*** **L. (Guttiferae)**

Common names
Eng.: Mangosteen
Beng.: Mangustan
Hindi: Mangustan

Distribution

Tree native to Malaysia, is now cultivated in lower slopes of Nilgiris.

Parts used: Fruits.

Pharmacological activities

Possess anti-inflammatory properties in Freund's complete adjuvant induced arthritis in animals.

Chemical constituents

Three new xanthones-gartanin, 8-deoxy- and normangostin-isolated from fruits; mangostin isolated and its structure confirmed; l,3,6,7-tetra hydroxycanthone and its glucoside isolated from heartwood; cyanidin-3-sophoroside and cyanidin-3 glucoside from rinds.

36. ***Glycyrrhiza glabra*** **L. (Papilionaceae)**

Common names
Eng.: Liquorice
Beng.: Jashtimadhu
Hindi: Mulhathi

Distribution

A perennial herb, native to the Mediterranean region and is now grown in Punjab, Jammu and Kashmir and S. India.

Parts used: Roots.

Pharmacological activities

Root extract was tested in formaldehyde externally induced arthritis, Freund's complete adjuvant induced arthritis, Pyrexia and cotton pellet induced granuloma in animals and its anti sesquiterpene inflammatory and antipyretic activities were confirmed.

Chemical constituents

See Antiulcer Chapter.

37. ***Hedychium coronarium*** **Koenig ex Retz. (Zingiberaceae)**

Common name
Ginger lily.

Distribution

An ornamental rhizomatous herb occurring throughout the moist parts of India.

Parts used: Rhizome.

Pharmacological activities

Essential oil, obtained from rhizome is active against gram positive bacteria and fungi. Powdered rhizomes are used in medicines as febrifuge, decoction is considered as anti-rheumatic and tonic.

Chemical constituents

Essential oil (0.1%) from rhizomes contained eucalyptol.

38. *Hibiscus vitifolius* L. (Malvaceae)

Common names
Beng.: Bankapas
Hindi: Bankapas

Distribution

A bushy shrub distributed in hotter parts of India.

Parts used: Seeds.

Pharmacological activities

Seeds are reported to possess powerful anti-inflammatory activity. Activity was tested in carrageenan in gartanin induced oedema, Granuloma pouch, Mediator induced oedema in animals.

Chemical constituents

A new gossypetin glucuronide-hibifolin isolated from flowers along with gossypin and characterized.

39. *Juniperus communis* Linn. (Pinaceae)

Common names
Eng.: Juniper
Beng.: Havusha
Hindi: Aaraar

Distribution

The plant is widely available in Northern part of India at the altitude of more than 1350 m.

Parts used: Leaves.

Pharmacological activities

Potent anti-inflammatory activity of leaf extract was confirmed in carrageenan induced paw oedema, granuloma pouch in rats. Yeast induced pyrexia in rat was significantly reduced by leaf extract.

Chemical constituents

Communic acid, m.p. 228° (0.12%) from bark of *J. communis* and longifolene, Juniperol, Beta-sitosterol, stigmasterol and disterpene phenol-totarol isolated from its bark. Production of essential oil and tropolone by plant tissue culture has been reported.

40. *Madhuca longifolia* (Koenig) Macb. (Sapotaceae)

Common names
Eng.: South Indian Mahua
Beng.: Rainjani
Assam: Awnapat
Hindi: Mohua

Distribution

Cultivated in U.P., Bihar, A.P., Karnataka, Bengal and Maharashtra.

Parts used: Barks.

Pharmacological activities

Plant extract showed anti-inflammatory activity in carrageenan induced oedema, formaldehyde induced arthritis in rats.

Chemical constituents

A new saponin-bassianin isolated which on hydrolysis yielded basic acid, glucose, arabinose, xylose and rhamnose; Beta-sitosterol-Beta-D-glucoside, stigmasterol and 3 Beta-caproxyolean-12-en-28-ol isolated from leaves; Beta-carotene, n-octacosanol, sitosterol, its Beta-D-glucoside, stigmasterol, 3 Beta palmitoxyolean-12-en-28-ol, oleanolic acid, quercetin, erythrodiol and palmitic acid isolated from leaves myricetin and its 3-O-L-rhamnoside isolated from leaves; structure elucidation of saponins A and B isolated from seeds; quercetin, myricetin-3-O-L-rhamnoside and quercitrin isolated; soil bacterial hydrolysis of saponins of seed kennels yielded proto bassic acid and prosapogenol; Mi-saponin A and Mi-saponin B isolated from seeds.

41. *Maminea longifolia* Planch. & Triana. (Clusiaceae)

Common names
Eng.: Alexandrian laurele
Beng.: Nagesar
Hindi: Nagkesar

Distribution

A tree cultivated in S. India.

Pharmacological activities

Plant possesses potent anti-inflammatory properties.

Chemical constituents

Squalene, cycloartenol compesterol, stigmasterol and Beta-sitosterol. Flowers contained vitexin and mesoinositol. Two new 4-alkylated coumarins-surangin A and B from roots.

42. *Mesua ferrea* L. (Clusiaceae)

Common names
Eng.: Iron wood
Beng.: Nagkeshar
Guj.: Nagchampa

Distribution

A tree found in eastern Himalayas, Assam, West Bengal, W. Ghats, Travancore and the Andaman Islands.

Parts used: Seeds.

Pharmacological activities

Seeds yield a fatty oil used as an embrocation in rheumatism. Anti-inflammatory activity was tested in carrageenan induced oedema cotton pellet induced granuloma and Granuloma pouch in animals.

Chemical constituents

Mammeisin isolated from seeds; a new 4-mesuagin isolated from seed oil and characterised; mammeigin and mesuol isolated from seed oil; a new biflavanone-mesuaferrone A-isolated from stamens and characterized as 8,8'-binaringenin; structure elucidation of another biflavone mesuaferrone B-isolated from stamens.

43. *Michelia champaca* Linn. (Magnoliaceae)

Common names
Beng.: Champa
Assam: Phulchopa
Bombay: Champa
Eng.: Champak

Distribution

A large tree, cultivated mainly in South India, West Assam and Bengal.

Parts used: Roots, root barks.

Pharmacological activities

The dried root and root bark mixed with curdled milk is useful as an application to abscesses, clearinga way or maturing the inflammation.

44. *Moringa oleifera* Lam. (Moringaceae)

Common names
Eng.: Drum stick tree
Beng.: Sagina
Hindi: Sahinjan, Soanjna

Distribution

A small tree native to India.

Parts used: Seeds.

Pharmacological activities

Oil obtained from seeds in used medicinally in gout and acute rheumatism.

Chemical constituents

See Anti-fertility Chapter.

45. *Myrtus communis* L. (Myrtaceae)

Common names
Eng.: Myrtle
Beng.: Sutrasowa
Hindi: Vilayti mehendi, Murad

Distribution

A small tree cultivated in N.W. India.

Parts used: Berries.

Pharmacological activities

Myrtle oil applied in rheumatism and considered as rubifacient.

Chemical constituents

Isolation and structure elucidation of two acylphloroglucinols A and B; limonene (23.4), linalool (22.2), a phenyl coumarin pinene (14.5), cineol (11.6), P-cymol (1.8), camphene (0.5), pinene(0.3%) and traces of car-3-ene found in leaf essential oil.

46. *Nyctanthes arbor*-tristis L. (Oleaceae)

Common names
Eng.: Tree of Sorrow, Night flowering jasmine
Beng.: Shephalika
Hindi: Harsinghar, seoli

Distribution

A large shrub or small tree grown as an ornamental.

Parts used: Leaves.

Pharmacological activities

Leaves used in rheumatism and fevers; decoction given in sciatica.

Chemical constituents

Flavonol glycosides-astragalin and nicotiflorin isolated from leaves. A new iridoid-nyctanthoside isolated and characterised; crocin-1 (Beta-digentiobioside ester of α-crocetin), and crocin-3 (Beta monogentio-bioside ester of α-crocetin) isolated from flowers; D-mannitol isolated from flowers; detection of astragalin and nicotiflorin by chromatography; a new glycoside naringenin-4'-O-Beta D-giucopyranosyl-a-xylopyranoside isolated from stem along with Beta sitosterol.

47. *Ocimum basilicum* L. (Lamiaceae)

Common names
Eng.: Common basil
Beng.: Bantulsi
Hindi: Babui tulsi

Distribution

Indigenous to the lower hills of Punjab cultivated throughout the greater part of India.

Parts used: Whole plant.

Pharmacological activities

Plant considered antipyretic agent.

Chemical constituents

See Antiulcer Chapter.

48. *Paederia scandens* Lour. (Rubiaceae)

Common names
Beng.: Gandha
Hindi: Ghandhali, Somaraji
Guj.: Gnadhana

Distribution

Found in C. and E. Himalayas extending to Calcutta.

Parts used: Leaves and stems.

Pharmacological activities

Plant extract showed anti-inflammatory activity stronger than that of acetylsalicylic acid and weaker than that of hydrocortisone.

Chemical constituents

Hentriacontan, hentriacontanol, methyl mercaptan, ceryl alcohol, palmitic acid, sitosterol, stigmasterol, ursolic acid and iridoid glycosides-asperuloside, paederoside and scandoside isolated from leaves and stems.

49. *Pinus roxburghii* Sarg. (Pinaceae)

Parts names
Eng.: Long leaved pine
Beng.: Pine
Hindi: Chir, Sarala

Distribution

Found in Eastern and Western Himalayas.

Parts used: Stem.

Pharmacological activities

Oil of terpentine obtained by purification. From oil of turpentine, oleoresin is used in rheumatic pains.

Chemical constituents

Oleoresin obtained from oil of turpentine, Friedelin, ceryl alcohol and Beta-sitosterol isolated from bark, hexacosyl ferulate isolated and structure confirmed by synthesis.

50. *Piper longum* L. (Piperaceae)

Common names
Eng.: Long pepper
Beng.: Pipul
Hindi: Pipal, Piplamul

Distribution

Native of India and cultivated in W. Ghats, Karnataka and Tamil Nadu to some extent in W. Bengal.

Parts used: Roots and fruits.

Pharmacological activities

Roots and fruits are used as counter-irritant and analgesic for muscular pain and inflammations.

Chemical constituents

Two new monocyclic sesquiterpenes, bp. 235° (15.5) and b. p. 247° (11 .1%) from essential oil; two new alkaloids-piper longuminine (piplartine), m.p. 124° and piperlonguminine-from roots and stem bark, characterized as N-(3,4,5-trimethoxy cinnamoyl)-piperidin-2-one-5-ene and isobutylarnide of piperic acid respectively; a new sesquiterpenic hydrocarbon containing tetrasubstituted double bond from essential oil; sesamin isolated; isolation of N-isobutyl deca-trans-2-trans-4-dienamide, m.p. 69°.

51. *Pluchea indica* Less. (Asteraceae)

Common names
Beng.: Kukronda, Manijhu rukha

Distribution

A shrub found in salty marshes in Sundarbans. Distributed in tropical and sub-tropical regions of Asia. It contains around 50 species out of which 6 have been recorded in India.

Parts used: Roots and leaves.

Pharmacological activities

Root extract possessed potent anti-inflammatory properties. Possible mechanisms of actions were established.

Chemical constituents

Endesmane derivatives of the enactahemone. Pluchea indica is known to contain 3-(2',3'-diacetoxy-2'-methylbutyryl)-cuanhtemone (molecular formulae $C_{24}H_{36}0_{8}$), col Parts odurless prism shaped crystals of m.p. 165°C. The plant is also known to contain linaloolapiosyl glucoside (colourless viscous oil, molecular formulae $C_{21}H_{3}0_{10}$), linalool glucoside (colourless viscous oil, molecular formulae $C_{16}H_{28}0_{6}$), 9-hydroxy-linalool glucoside (colorless viscous oil, molecular formulae $C_{15}H_{28}0_{7}$), plucheoside A (amorphous powder and B. Plucheoside B (amorphous powder).

52. *Pluchea lanceolata* C.B. Clarke. (Asteraceae)

Common names
Punjab: Sarmei
Hindi: Rasna, Marmandai
Sans.: Rasna

Distribution

A small shrub found commonly in Punjab and U.P.

Parts used: Whole plant.

Pharmacological activities

Plant used in rheumatoid arthritis.

53. *Psoralea corylifolia* Linn. (Fabaceae)

Common names
Eng.: Bakuchi
Beng.: Barachi
Hindi: Babchi

Distribution

A common herb, available in India.

Parts used: Seeds.

Pharmacological activities

Seeds are specially recommended for inflammatory diseases of skin.

Chemical constituents

Isolation and separation of psoralen and isopsoralen from seeds; isotatioii and separation of psoralen and from seeds; a new isoflavone-neobava isoflavone- and a new chronenochalcone-bavachromene isolated from seeds and fruits; psoralidin, 4′-O-methylbavachalcone, 7-0-methylbavachin and isobavachalene isolated from seeds; a nevel monoterpene phenol-(+)bakuchiol from seeds; its structure and absolute configuration [(S)-chirality] assigned; a new isoflavone-corylin-isolated and charactensed as 7-hydroxy-6", 6"-dimethylpyrano-(2",3", 4′,3′)-isoflavone; partial synthesis of corylin; high yield of psoralen and angelicin found in seeds; new coumestrol-corylidin isolation from seeds together with triacontane and sitosterol-Beta-D-glucoside; new formylated chalcone-neobavachalcone-isolated from seeds and its synthesis; a new isoflavone-carylinal isolated from seeds together with neobavaiso-flavone as their methyl ethers; isolation and structure elucidation of another isoflavone psoralenol from seeds.

54. *Ranunculus aquatilis* Linn. (Ranunculaceae)

Common names

Eng.: Water crow foot, water Fennel
Kashmir: Tohiub

Distribution

Punjab Plain, W. Himalaya from India to Kumaon, Baluchisthan, Afghanistan, N. Africa and Europe.

Parts used: Leaves.

Pharmacological activities

The leaves are applied as bilister to the wrists in rheumatism.

55. *Ranunculus arvensis* Linn. (Ranunculaceae)

Common names

Eng.: Corn Butter cup, Crows-claws
Punjab: Chambul

Distribution

Western Himalayas from Kashmirro Kumaon, Mt. Abu, Afganisthan, Europe, N. Africa.

Pharmacological activities: In Europe the plant is used in intermittent fevers and gout.

56. *Ranunculus muricatus* Linn. (Ranunculaceae)

Syn. ***R. cabulicus*** Boiss.

Common name

Eng.: Field crowfoot

Distribution

Punjab, Himalayas, Punjab-Kashmir, W. Africa, Europe and temperate N. America.

Pharmacological activities

In Europe the plant is used in intermittent fevers, gout.

57. *Ricinus communis* L. (Euphorbiaceae)

Common names

Eng.: Castor, Castor oil plant
Beng.: Bheranda
Hindi: Erandi

Distribution

A small tree cultivated chiefly in Andhra Pradesh, Maharashtra, Karnataka and Orissa. It is also found as wild.

Parts used: Roots.

Pharmacological activities: Decoction of roots is useful in lumbago.

Chemical constituents

See Antidiabetic Chapter.

58. ***Sagittaria sagittifolia*** **L. (Alismataceae)**

Common names
Eng.: Old-world arrowhead
Beng.: Muya muya
Hindi: Chotokut

Distribution

A herb found throughout India.

Parts used: Leaves.

Pharmacological activities

The leaves are used in sore throat and inflammation of breast.

Chemical constituents

Hentriacontane and sitosterol isolated; a diterpene-sagitariol isolated and characterized as labda-7, 1 4-diene-13 (5), 17-diol.

59. ***Semecarpus anacardium*** **L. (Anacardiaceae)**

Common names
Eng.: Marking-nut tree
Beng.: Bhela
Hindi: Bhilawa

Distribution

A tree occurring in hotter parts of India.

Parts used: Fruits.

Pharmacological activities

Fruits used for rheumatism.

Chemical constituents

Bhilawanol from fruits was found to be a mixture of 1, 2-dihydroxy-3-(pentadecen lyl-8′)-benzene-l,2-dihydroxy 3-(pentadecenlyl-8′, 11′)-benzene; studies on methylated bhilawanol showed that it contained more than seven components; two major components identified as dimethyl ethers of i-pentadeca-8-enyl-2,3-dihydroxy-benzene (I) and I -pentadeca-7, 10-dienyl- I,3-dihydroxy benzene (II); defatted nuts yielded three biflavones A, B, and C; latter two compounds characterized as 3′8-binaringenin and 3′,8-biliquiritigenin; reexamination of bhilawanol showed into be comprised of two components, 1 ,2-dihydroxy-3-pentadecenyl benzene (32–32%) and its corresponding diene analogue (68–70%); a new B-biflavan tetrahydrobusta-flavone and tetra hydro-amento flavone isolated from nuts; leaves yielded only amentoflavone.

60. ***Sida acuta*** **Burm. (Malvaceae)**

Common names
Beng.: Sweet berela
Hindi: Bariara, kharenti
Mar.: Tupharia

Distribution

A common under shrub.

Parts used: Leaves and roots.

Pharmacological activities

Leaves used in rheumatic affections. Roots are used as antipyretic.

Chemical constituents

Ecdysterone isolated from the plant.

61. *Tinospora crispa* Linn. (Menispermaceae)

Common names
Hindi: Gulancha
Beng.: Goloncha

Distribution

A climbing shrub, found throughout the tropical parts of India known by the same regional names as *Tinospora cordifolia* and used medicinally.

Parts used: Leaves.

Pharmacological activities

In the Philippines islands it is considered to be a panacea to be applied to all bodily afflictions. It is given in chronic rheumatism.

Chemical constituents

Sodium, potassium, calcium, iron, aluminium, copper and zinc estimated in leaves; two unidentified alkaloids, one hydroxy compound, m. p. 950, y-sitosterol and sitosterol and essential oil, b. p. 116°, isolated from leaves.

62. *Tinospora malabarica* Lam. (Menispermaceae)

Common names
Beng.: Podma
Hindi: Gulancha, Gurch
Oriya:Gulochi, Guduchi

Distribution

Bengal, Assam, Khasiam Orissa, Kankan, Kanara and nearly all districts of Tamil Nadu.

Parts used: Leaves and stems.

Pharmacological activities

In China and Tong king the fresh leaves and the stems are used in the treatment of chronic rheumatism, fumigations are recommended in piles and ulcerated wounds.

63. *Tylophora asthmatica* W. & A. (Asclepiadaceae)

Common names
Oriya: Mendi
Beng.: Anantämul
Hindi: Antamul

Distribution

A climber distributed in Assam, W. Bengal, Orissa and Peninsular India.

Parts used: Whole plant.

Pharmacological activities

Plant extract was used in different animals models like carrageenan induced oedema, cotton pellet induced granuloma and granuloma pouch and found significant anti-inflammatory activity.

Chemical constituents

Three alkaloids A, B and C isolated; alkaloids B and C characterized as desmethyl-tylophorine and desmethyl-thylophorinine respectively; crystal structure of tylophorinidine and relative stereo chemistry of tylophorinine and tylophorinidine; (+) septicine and (+) isotylocrebrine also isolated from fresh leaves; dehydrotylophorine, anhydro dehydrotylophorinidine and anhydro dehydrotylophorinidine isolated: stero-chemistry of tylophorinine and tylophorinidine; absolute configuration of tylophorine; tylophorine and tylophorinine content in leaves is a function of plant growth phase and is highest during flowering period; Gama-fagarine and skimmianine isolated from roots and aerial parts.

❑❑❑

3

MEDICINAL PLANTS FOR ANTIULCER

INTRODUCTION

Peptic Ulcer

An ulcer is defined as a breach in the continuity of the epithelial lining of more than 5 mm in diameter, with associated inflammation.

Peptic ulcer disease is an important cause of morbidity and health care costs; estimates of expenditures related to work loss, hospitalization and out patient care (excluding medication costs) are more than $5.6 billion per year in the United States. In the United States, approximately 4 million people have peptic ulcer (duodenal and gastric), and 350,000 new cases are diagnosed in each year. Around 100,000 patients are hospitalized yearly, and about die each year as a result of peptic ulcer disease. The peptic ulcer are remitting, relapsing lesion that are most often diagnosed in middle age to older adult, but they must first become evident in young adult life. Although the prevalence of peptic ulcer is decreasing in many Western communities, it still affects approximately 10% of all adults at some time in their lives.

Types of Ulcer

There are two main type of peptic ulcer.

1. Gastric ulcer
2. Duodenal ulcer.

The description of both type shown in Table 1.

Table 1: Difference in duodenal ulcer and peptic ulcer

	Duodenal ulcer	*Gastric ulcer*
Location	Anterior wall of duodenum	Located along the greater curvature
Male to female ratio	3:1	1.5 to 2 : 1
Hyperacidity	Only a minority of patient	Less common
H. pylori infection	Present in all patients	70 % of patients
Chronic gastritis	85 to 100% of patients	65% of patients
Parietal cell mass (PMC)	Decreased parietal cell mass	Increased parietal cell mass
Gastric emptying rate	Delayed gastric emptying	Increased gastric emptying

Aggressive Factor and Defensive Factor

Peptic ulcer due to imbalance in the defensive and aggressive factor.

These are as follows:

1. Aggressive Factor

i. Gastric acid and pepsin

Acid and pepsin are the major aggressive factors believed to play a role in the pathogenesis of peptic ulcer disease. On the average, patients with duodenal ulcer disease have higher than normal basal, nocturnal, and pentagastrin-stimulated acid secretion.

ii. Helicobacter pylori

H. pylori is a non-sporing, curvilinear gram negative rod measuring approximately 3.5 – 0.5 µm.*H. pylorus* is established as a cause of peptic ulcer disease. *H. pylori* infection causes the majority of duodenal (90–95%) and gastric ulcers (60–70%). In the industrialized world the prevalence of *H. pylori* infection in the general population rises steadily with age, and in the UK approximately 50% of those over the age of 50 years are infected. In many parts of the underdeveloped world infection is much more common and is often acquired in childhood.

iii. Cigarette smoking

Smoking is to known to have several adverse effects on the upper gastrointestinal tract. Smoking confers an increased risk of gastric ulcer and to a lesser extent, duodenal ulcer. The particular interest for ulcer perforation is the finding that smoking cause immediate vasoconstriction in the mucosa, ischemia reduces mucosal resistance. Smoker have a three fold higher mortality from peptic ulcer than non- smoker. The prevalence of peptic ulcer in smokers (26.87%) is double that of non smokers (13.38%). Cigarette smoke extract significantly reduced mucus synthesis in animal with or without ulcer.

iv. Alcohol

There is little evidence of an association between alcohol use and peptic ulcer. But alcoholic cirrhosis is associated with an increased evidence of peptic ulcer. Substance present in the alcoholic beverage stimulates gastric acid secretion *via* M3 receptor, muscarinic M1 receptor, H2 receptor and CCK2 receptor on parietal or entero chromaffinecells.

v. Stress

The link between stress and peptic ulcer. Psychological stress is not only empirically associated with ulcers, but is a very plausible risk factor for ulcer disease. Under stress, the amount of acid reaching the duodenum may increase further because gastric motility has changed or meals have been missed; sleep less, thereby increasing their susceptibility to ulcer by mechanisms that are not related to acidity. Stress induces activation of sympathetic and parasympathetic nervous system causing vascular and smooth muscle contraction of the stomach, leading to ischemia due to reduced blood flow. Ischemic stomach generate reactive oxygen species (ROS) especially OH for oxidative damage of the mucosa.

vi. Physical activity

The relative risk of having a duodenal ulcer was greater in those with a high level of occupational activity than in those undertaking sedentary work.

2. Environmental Factors

Researchers have been shown that the occurrence of peptic ulcer in winter is related to the higher air pressure and temperature. The high occurrence of PU in summer is likely to be related to the lower air pressure. Acute, sub-acute and slow stresses on the human body are likely the cause of PU in winter and spring.

3. Protective Factors

i. Mucus

A layer of water-insoluble mucus gel has been shown to form a continuous cover over the gastro-duodenal mucosal surfaces, of median thickness of 180 micron in stomach in humans. This adherent mucus is the first line in mucosal defense against the natural aggressors, acid and pepsin, in the lumen. Mucus gel is a diffusion barrier to pepsin in the lumen, preventing proteolysis of the underlying epithelial cells.

ii. Bicarbonate

The primary function of the mucosal bicarbonate secretion is to neutralize acid diffusing into the mucus gel layer and to be quantitatively sufficient to maintain a near-neutral pH at the mucus-mucosal surface interface. Mucus gel traps bicarbonate secreted by the epithelium and therefore acts as a layer in which luminal acid that diffuses towards the epithelium is neutralized. Sensory afferent nerves, sensing acid back-diffusion into the lamina propria, trigger an increase in epithelial bicarbonate secretion.

iii. Surface active phospholipids

The luminal surface of the stomach of a layer of surfactant-like molecules that rendered the mucosal surface hydrophobic, and therefore resistant to damage induced by luminal acid. Recently reported that non steroidal anti-inflammatory drugs (NSAIDs) pre associated with zwitter ionic phospholipids have markedly lower gastrointestinal toxicity following oral administration than standard NSAIDs. Prostaglandin E, is also believed to increase the surface hydrophobicity of gastric mucosa by increasing surface-active phospholipids surfactant.

iv. The epithelium

Intracellular tight junctions provide a barrier to the back diffusion of hydrogen ions. Epithelial disruption is followed rapidly by restitution, in which existing cells migrate along the exposed basement membrane to fill in the defect and restore epithelial barrier integrity.

v. Blood flow

Underlying the surface epithelium of the stomach is a dense network of capillaries.

Blood flow contributes to protection by supplying the mucosa with oxygen and HCO_3-, and by removing H+ and toxic agents diffusing from the lumen into the mucosa. Low mucosal blood flow predisposes to injury, whereas high blood flow protects against injurious agents.An increased blood circulation with increased oxygen supply to the ulcer margin is important for rapid ulcer healing.

vi. Restitution

It is likely that damage to the gastric epithelium occurs on a daily basis. The term restitution refers to this epithelial repair process, which involves rapid migration of healthy cells from the gastric pits over the denuded basement membrane.

Epidermal growth factor (EGF), transforming growth factor- (TGF-), hepatocyte growth factor (HGF), and trefoil factors (TFFs) are mainly involved in the reconstitution of the epithelial structures.

vii. Prostaglandin

Prostaglandin (PG) acts as a natural inhibitor of acid secretion by modulating the activity of the adenyl cyclase and cAMP formation *via* the inhibitory CTP- binding (Gi) protein.

PGs of the E and I series and TXA2 are generated by the gastrointestinal mucosa and released into the lumen upon neural or hormonal stimulation; they probably participate in the maintenance of mucosal integrity and micro circulation. Exogenous PGs of the E and I series inhibit gastric acid secretion and stimulate alkaline secretion while increasing mucosal blood flow. All PGs, including those non inhibitory for acid secretion, are cyto protective against various ulcerogens and necrotizing agents.

viii. Nitric oxide

Nitric oxide synthase (NOS) also exists in constitutive and inducible forms. In the stomach, constitutive forms include that found in the endothelium (eNOS) and in enteric neurons (nNOS). When appropriately stimulated, such as after exposure to endotoxin, inducible NOS (iNOS) can also be detected. It has been suggested that nitric oxide derived from iNOS can exert cytoprotective effects in the stomach. NO can prevent mucosal injury in the stomach may be by gastric blood flow. Endogenous NO plays an important role in the maintenance of blood flow around the ulcer, in the angiogenesis in the granulation tissue and thus, in the healing of gastric ulcers.

Role of Reactive Oxygen Species in Peptic Ulcer Disease

Recently, interest has been focused on the role of ROS in gastro-duodenal pathogenesis related to gastric hyper secretion and gastro duodenal mucosal damage.

Central Peptidergic Control of Gastric Acid Secretion

The description of inhibition and Stimulator shown in Table 2.

Pharmacology of the Digestive System

The Salivary Glands

Saliva contains an amylase (ptyalin) which passes with the food into the stomach where it begins the digestion of starch. Saliva serves

Table 2: Peptide that affects gastric acid secretion after central injection

Inhibition	*Stimulators*
Corticotropin releasing factor	Thyrotropin realeasing hormone
B endorphine	Somatostatin
Bombesin	Neuropeptide Y
Neurotensin	Galanin
Calcitonin	
Calcitonin and calcitonin gene related peptide	
Interleukin-1	

a number of other useful functions: by keeping the mouth and lips moist, it aids articulation and by moistening and lubricating the food and facilitates swallowing. Cessation of salivary secretion causes the sensation of thirst which signals a fall in the body's water content.

In the normal subject, the arrival of a bolus of food at the first part of the esophagus initiates a peristaltic wave which carries it towards the cardiac sphincter. Coordinated reflex relaxation of the cardiac sphincter allows the peristaltic wave to carry the bolus into the stomach. Here, churning movements and peristaltic waves ensure that the food circulates and becomes thoroughly mixed up with the gastric secretions, pepsin, hydrochloric acid (HCI) and mucus.

The constituents of gastric juice, necessary for digestion are pepsin (a proteolytic enzyme) and hydrochloric acid which together initiate the digestion of protein. Salivary amylase also operates in the stomach until hydrochloric acid penetrates the food mass and arrests amylase activity. Digestion of protein and starch continues, and that of other substances commences in the duodenum, which receives the pancreatic secretions. In the rest of the small intestine, success entericus is secreted.

Hydrochloric acid is produced by the oxyntic (or parietal) cells of the stomach. These cells also secrete Castle's intrinsic factor, the substance necessary for the absorption of vitamin B12. Hydrogen ions, derived from water, are secreted by the gastric glands against a concentration gradient using energy obtained from a series of oxidative reaction. The hydroxyl ions combine with carbon dioxide under the influence of carbonic anhydride and the resulting bicarbonate ions pass into the blood, following ion exchange with chloride ions, which accompany the hydrogen ions into the glandular lumen: Therefore, this separation of the hydrogen and hydroxyl ions, results in the production of hydrochloric acid in the stomach and of bicarbonate in the blood. It may also be mentioned in this regard that more vigorous the secretion of acid the more alkaline will be the blood leaving the stomach, which is due to the 'alkaline tide' in the blood which accompanies gastric digestion.

Secretion of gastric, which controls the production of pepsinate and HCI, is stimulated by the presence of food within the lumen, and is inhibited by the presence of increase levels of HCI in the gastric tantrum. Hence, the secretary activity of the stomach in turn depends on the quantity of food within the lumen.

Peristaltic waves propel the food through the pylorus, which periodically relaxes synchronously with the stomach, thereby permitting the chime to enter the duodenum. Co-ordination of gastric and duodenal motility ensures the one-way flow of chyme from the stomach through the pylorus.

At this stage, two other hormones, secret in and cholecystokinin pancreozymin (CCKPZ) are secreted by the upper small intestine. Secretin inhibits gastric secretion and stimulates the pancreas to produce a profuse watery fluid with a high bicarbonate content, which passes into the duodenum and is necessary for neutralization of the residual gastric HCI and inactivation of pepsin, thus protecting the duodenal mucosa from damage by gastric acids.

CCKPZ lowers the muscle tone of the cardiac, stimulates the secretion of pancreatic fluid rich in digestive enzymes, causing the gall bladder to contract. As a result, a mixture of bile and pancreatic juice enters the duodenum and the process of digestion continues in an alkaline medium.

The Digestive System

The process of digestion causes breakdown of food material by chemical action and to convert them into more simple forms. These simpler forms can be absorbed into the blood and utilized by the various tissues of the body according to their requirements.

The process of digestion takes place in the alimentary canal. It is aided by certain accessory organs: the salivary glands, liver and pancreas. The digestive process can be described in four stages:

(a) Ingestion
(b) Digestion
(c) Absorption
(d) Excretion

Anatomy

The human digestive tract is comprised of the following parts:

- Mouth
- Pharynx
- Oesophagus
- Stomach
- Small Intestine – Duodenum
  – Jejunum
  – Ileum
- Large Intestine
- Rectum
- Anus

The Mouth

The mouth is the upper portion of the alimentary canal.

The buccal cavity is found between the inner walls of the cheeks and the gums which contain the teeth. The salivary glands which open into the buccal cavity are

(a) Parotid
(b) Submandibular
(c) Sublingual

Saliva has a pH 6.7.

Pharynx

The pharynx is a musculo membranous tube whose constricted part ends in the oesophagus.

Oesophagus

The esophagus is a muscular tube about 10 inches (25 cm.) long. It extends from the pharynx and transports the bolus of food to the stomach by a series of peristaltic movements.

Stomach

The stomach is a dilated portion of the alimentary canal. Shape of the normal stomach is generally like the letter 'J'. The capacity of the average stomach is about 1.12 to 1 .70 litres.

The stomach consists of the following parts

a. The opening at the upper end known as the cardiac orifice, where the oesophagus meets the stomach.
b. The Fundus, the upper part which in general contains a bubble of air.
c. The body, constituting the main part of the stomach.
d. The antrum, forms the lower part of the 'J' and leads to the pyloric sphincter which

separates the stomach from the duodenum.

The stomach-wall consists of four coats:

a. An outer peritoneal coat, which is a serious membrane and forms a part of the peritoneum.
b. A muscular coat having three layers:
   I. Longitudinal fibers
   Il. Circular fibers
   Ill. Oblique fibers
c. A sub mucous coat containing blood vessels and lymphatic.
d. A mucous coat which is lined by columnar epithelium and containing numerous lymphatic's. All the cells secrete mucus. The inner surface is covered by tiny ducts of the gastric glands. The epithelium of the secreting part of the gland differs in various parts of the stomach.

Cardiac Glands

The gland is located near the esophageal end. These are tubular glands, either simple of branched and secrete an alkaline mucus.

Glands of the Fundus

These are tubular glands containing different types of cells:

Peptic cells - Produce pepsin
Oxyntic cells - Produce hydrochloric acid
Mucus neck cells - Produce mucin

Pyloric Gland

The glands in the pyloric canal are also tubular in character. They produce mainly alkaline mucus.

The secretion from the body and the fundus glands is rich in acid and enzymes, whereas the secretion of the pylorus is alkaline, rich in mucus and poor in enzymes.

Small Intestine

The small intestine is the portion of the alimentary tract extending from the pyloric end of the stomach and continues till the large intestine (the caecum). It is about six metres long and consists of three parts:

(i) The Duodenum
(ii) The Jejunum
(iii) The ileum

The Duodenum - the first ten inches of the small intestine is shaped like a horse shoe, the curve encircling the head of the pancreas. The bile ducts and the pancreatic duct open into the ampulla of Vater (ampulla Vateri).

The Jejunum occupies the upper two fifths of the remaining small intestine.

The ileum occupies the rest of the small intestine.

The wall of the small intestine, like the stomach wall has also four coats, *viz.*:

1. Serous coat (part of the peritoneum)
2. Muscular coat
3. Sub mucous coat
4. Mucous Membrane (inner membrane)

As in the stomach, the mucous membrane is arranged in folds. These folds increase the surface area from which secretion and absorption can take place. The surface of the mucous membrane is slightly roughened by minute projections called VILLI. Each villus contains capillaries into which the end products of food are absorbed and a central lymphatic vessel (lacteal) where fats are absorbed.

The mucous membrane of the small intestine contains glands which secrete the intestinal juice (SUCCUS ENTERICUS).

Large Intestine

The large intestine extends from the end of the ileum to the anus.

Diseases of the Stomach and Duodenum

Peptic ulcer

In the western hemisphere, peptic ulcer is one of the commonest diseases of the alimentary tract. It affects particularly the working years of a patient's life and its social implications are therefore considerable and this disease is a perennial problem encountered by the clinicians around the world.

Aetiology

The term 'peptic ulcer' refers to an ulcer found in the lower end of the esophagus, the stomach, the duodenum, in the small intestine after surgical anatomosis to the stomach, or rarely at the junction of a Meckel's diverticulum with the small intestine. Although the immediate cause of peptic ulceration is digestion of mucosa by acid and pepsin of the gastric juice, the sequence of events that leads to the development of the ulcer are unknown. Digestion by acid-pepsin cannot be the only factor involved, since ulcers do not develop in some normal people who secrete acid and pepsin in substantial amounts. The problem is aptly stated in the question: 'Why does the stomach not digest itself?' The answer lies on the fact that normal stomach is capable of resisting digestion by its own secretions, so that the problem of ulcer aetiology may be written as:

Acid plus Pepsin *versus* Mucosal Resistance:

Theoretically, peptic ulceration may be caused due to either increase in gastric secretion or if the resistance of the mucosa is reduced. It is convenient to consider the factors believed to be involved in the development of chronic peptic ulcers according to how they may alter either side of the balance.

1. Gastric Hyper Secretion

There is strong evidence to indicate that ulcers occuronly in the presence of acid and pepsin. Peptic ulceration has never been found in patients with pernicious anemia, and it is very doubtful if a benign chronic ulcer ever occurs in association with achlarhydria. On the other hand, severe intractable peptic ulcer occurs in patients affected with the Zollinger-ellison syndrome, which is characterized by very increased gastric secretion. Moreover, patients with duodenal ulcer secrete about twice as much acid as compared to normal individuals, and duodenal ulcer heals when gastric hyper secretion is reduced by vagotomy. However, gastric hyper secretion does not surface in all patients with duodenal ulcer. Thus, factors affecting mucosal resistance must be considered in patients with gastric ulceration.

2. Mucus and cellular turnover

Alkaline mucus is the first line of defense for the gastric mucosa, and it may act simply as a protective barrier. Moreover, the superficial layer of the gastric mucosa renews itself every two or three days and such minor damages in the mucosa are rapidly healed under normal circumstances alteration in the rate of cell renewal in the upper alimentary mucosa may account for the progression of ulcers from acute to chronic stage.

3. *Inflammatory change*

Gastritis, which may be both extensive and severe, commonly occurs in patients with gastric ulcer and since there is no definite evidence, it seems likely that the gastritis in fact precedes the development of the ulcer; if so, the inflammatory change may well predispose the development of ulceration by altering the resistance of the mucosa from digestion by the gastric secretion.

4. *Bile Reflux*

Reflux of bile and other secretions into the stomach occurs more frequently in patients with gastric ulcers than in normal individuals and in patients with duodenal ulcers, this may be attributed to an abnormality of pyloric sphincter. Irrespective of the mechanism, the increased concentration of bile salts in the stomach interfere with the integrity of the gastric mucosa, thus predisposing to the development of ulceration.

5. *Blood supply*

One of the classical theories of ulcer formation is that the resistance of the mucosa to digestion is reduced as a result of impaired blood supply; such impairment could occur either as a result of venous or arterial thrombosis, or due to 'shunting' of blood within the mucosa. Although it is unlikely that this is the cause of peptic ulceration in general but the same may account for the extensive ulceration that occurs in the elderly.

Apart from these local mechanisms, several general or constitutional factors are known to be associated with the development of peptic ulcer.

6. *Heredity*

There is no doubt that peptic ulcer tends to occur in families, and the tendency tends to run true to type, so that the children of parents with duodenal ulcer develop duodenal ulcer, and like wise for gastric ulcer. A strong family history is frequently found in patients who develop ulcer in childhood or adolescence.

There is an association between duodenal ulcer and blood group 0.

7. *Sex incidence*

There are obvious differences both in the incidence and behavior of peptic ulcer. Duodenal ulcer occurs 5 to 10 times more often in men than in women, and perforation occurs 20 times more often in men. Although these effects may be due to differences in the life pattern of men and women, there are grounds for supposing that the female sex hormones in some way protect against peptic ulcer. Women appear to be particularly protected against ulceration during pregnancy, when active ulcers are virtually unknown; ulcer symptoms which remit during pregnancy commonly recur soon afterwards. The incidence of ulcer symptoms and ulcer complications increase sharply in women at about the menopause. The sex difference in the incidence and behavior of gastric ulcer is much less than for duodenal ulcer.

8. *Environmental factors*

There are marked variations in the incidence of gastric and duodenal ulcer between different countries, and even between different parts of the same country *e.g.*, the proportion of duodenal ulcer to gastric ulcer is very much higher in Scotland than it is in London. There are also differences in the incidence of ulcer as between social classes so that duodenal ulcers tend to be evenly distributed throughout the entire population, while gastric ulcer occurs more commonly amongst poor people, especially in Britain.

9. Stress

There is an undoubted association between the development of acute ulcers in the stomach or duodenum and physical or mental trauma or surgical operations; similarly there is little doubt that acute anxiety is a significant factor in precipitating ulcer recurrences and sometimes ulcer complications such as haemorrhage or perforation. Although these effects are presumably mediated *via* the central nervous system, how stress or anxiety influences the occurrence or behavior of peptic ulcer is not known.

10. Seasonal factor

There are strong seasonal factors which can be shown to influence the incidence of ulcer symptoms and mortality rates; in Britain, ulcer mortality is at its lowest in August and September and begins to rise in October, with a further exacerbation of ulcer symptoms in the spring. The cause of these seasonal variations is unknown.

11. Other factors

The incidence of peptic ulceration appears to be increased in patients with chronic lung disease, chronic liver diseases and hyperparathyroidism; similarly, anti-inflammatory drugs such as the salicylates, phenylbutazone and indomethacin have all been associated with an increased incidence of peptic ulcer, especially of gastric ulcer. Corticosteroids are also believed to cause peptic ulcer.

Pathology

Ulcers occurring in the stomach and duodenum may be acute or chronic, the difference being that a chronic ulcer penetrates the muscular is mucosa, whereas an acute ulcer or an erosion does not. Chronic ulcers occur with remarkable regularity in certain sites: in the stomach on the lesser curvature just above the angulus or less frequently at or near the pylorus, while duodenal ulcers occur within 1 cm of the pylorus on the anterior or posterior wall. Acute lesions are frequently multiple, and are less regularly distributed. Benign ulcers occur only rarely on the greater curvature or on the anterior wall of the stomach.

Clinical Features

While there are good grounds for believing that gastric and duodenal ulcers are different diseases with different a etiology and natural history, it is convenient to describe the general features of 'peptic ulcer' as inclusive of both, noting differences where they occur. Peptic ulcer may present in different ways. The commonest presentation is of chronic, episodic dyspepsia extending over months or years. However, the ulcer may come to attention as an acute episode with bleeding or perforation, with little or no previous history. Occasionally the patient presents with the symptoms of gastric outlet obstruction, having had negligible dyspepsia previously.

1. Pain

This is the characteristic symptom of peptic ulcer, and it has three remarkable features - sharp localization to the epigastrium, relationship to food, and periodicity. Ulcer pain is typically referred to the epigastriurn, in the midline or to the right; wherever it occurs, it is usually sharply localized so that the patient can point to the site. This feature of ulcer pain is so striking as to be called 'the pointing sign'. Occasionally ulcer pain is not clearly localized; it may be referred diffusely in the epigastrium, the lower chest or to the back in the interscapular region in the fifth to eighth thoracic segments. Pain referred to the inters capular area, especially if it is a new feature, suggests the possibility that the ulcer has penetrated posteriorly, involving structures such as the pancreas. The description of the pain is not especially helpful, although patients commonly describe it as gnawing or burning. Pain varies considerably in severity, and it is sometimes helpful to ask the patient to qualify the symptom as 'pain' or as 'discomfort' as a measure of its intensity.

Most patients recognize a relationship of the pain to food, although the relationship varies between patients and in the same patient from time to time. Duodenal ulcer-pain tend to occur between meal times, so that the patient may describe it as 'hunger' pain, which is characteristically relieved by food. A notable feature of duodenal ulcer is pain awakening the patient from sleep between 2 and 4 am. The pain of gastric ulcer occurs less regularly; it frequently occurs within an hour of eating, is less often relieved by food, and it rarely occurs at night. Besides the characteristic relief obtained after eating, ulcer pain is almost invariably relieved by antacids, by vomiting and by bed rest in hospital.

Ulcer pain is characteristically episodic occurring regularly each day for days or weeks at a time, then disappearing, to recur weeks or months later. Between attacks, the patient feels perfectly well, and may eat and drink with impunity. Bouts of pain may at first last only a day or so at a time, and occur only once or twice a year. As the natural history evolves, however, episodes begin to last longer and occur more frequently, so that in severe cases remissions of pain may be short-lived and pain or discomfort becomes more or less persistent. The cause for these relapses is difficult to establish. Seasonal factors may be operative, sometimes psychological stress may be blamed, sometimes dietary indiscretion, and sometimes alcoholic excess. Most commonly, no reason can be found for the relapse.

Pain is sometimes absent or so slight as to be dismissed by the patient. Such individuals may complain of other symptoms such as a feeling of 'distension' in the epigastrium or a poorly defined sense of unease after eating. Other complaints include episodic nausea and sometimes anorexia, as well as heart burn or water-brash. Nausea, anorexia, vomiting and weight loss occur more frequently in gastric ulcer than in duodenal ulcer. Vomiting in ulcer patients almost always relieves pain and when it is persistent may result in weight loss. This helps to distinguish it from vomiting of psychological origin, in which weight is usually maintained. Persistent vomiting in an ulcer subject usually indicates some degree of gastric out flow obstruction, either due to spasm or organic narrowing of the gastric outlet. In such patients, vomiting is usually copious, so that the patient is 'surprised' at the volume; the patient often recognizes food eaten twelve or more hours previously, and he is aware of the unpleasant smell of the vomitus. Although there is no constant change in bowel rhythm during an ulcer relapse, some patients are aware of constipation or diarrhea when dyspepsia reappears.

2. Physical signs

The only physical sign that may be present is 'the pointing sign' by which the patient accurately indicates the site of pain; when accompanied by localized tenderness the sign is practically diagnostic of an ulcer. However, tenderness may be completely absent. In patients with gastric outlet obstruction, the stomach may be visibly distended, a succession splash may be present, and gastric peristalsis may be seen.

Diagnosis

The diagnosis of peptic ulcer can usually be made from the characteristic history, but it requires confirmation by barium meal examination or endoscopy. In some patients the peptic ulcer may present with bleeding, perforation or even pyloric stenosis with few or no previous symptoms. The patient with troublesome dyspepsia, in whom barium meal examination is negative, presents difficulties. Some of these patients will be found to have an ulcer at endoscopy and indeed some may be found to have an early carcinoma. Some clinical judgment has to be exercised in order to distinguish between dyspepsia of psychological origin and that due to organic lesions which have not been detected radiologically. In general, the history in the former is more diffuse, the symptoms rarely as clear cut, the pain poorly localized, and there are no sharply defined tender areas. Positive evidence of a psychological abnormality can usually be obtained. It is a good rule that patients who develop dyspepsia for the first time in middle age should be examined endoscopically if the X-ray is negative. Probably all patients with gastric ulcer should be examined endoscopically since the lesion can be inspected more closely and biopsies taken to exclude a carcinoma. This procedure is also indicated if the physician has any doubt about the nature of the disorder. Gastric secretion test: Although this test is of limited diagnostic value, a high acid output tends to support the diagnosis of duodenal ulcer and sustains a decision to operate; a low acid output is more consistent with a gastric than a duodenal ulcer.

1. Occult blood

The faeces should be tested for the presence or absence of occult blood in all patients with gastrointestinal symptoms. The test may be carried out of a smear of the stool obtained on rectal examination.

Treatments

There are three objectives in the management of peptic ulcer, namely, the alleviation of symptoms, the healing of the ulcer and the prevention of its recurrence. While there are effective methods for relieving symptoms and for accelerating ulcer healing, at present we have no means of preventing ulcer recurrence nor of altering the natural tendency of ulcer symptoms to remit and relapse. A major problem in interpreting the effectiveness of any particular treatment is the fact that ulcer symptoms do not necessarily reflect ulcer activity. Relief of symptoms can occur in a few days, whereas healing may take weeks. Thus the relief of symptoms, which is naturally the main concern of the patient, does not necessarily mean that the ulcer has healed; some ulcers may persist for many months without any further recurrence of pain, and without any change in size. The more chronic the ulcer, the less chance there is of obtaining healing, because the tissues are distorted by fibrosis. It seems a reasonable aim, therefore, to try to produce ulcer healing in the early stages, in the hope of preventing irreversible changes. Unfortunately, this ideal is difficult to attain for economic reasons, since the patient is frequently unable to take sufficient time off-work, and in any case admission to hospital cannot usually be offered to all patients with an active ulcer, but has to be reserved for patients with complications such as stenosis, bleeding or intractable pain.

1. Rest

The single most effective measure for the relief of ulcer pain and the promotion of healing is bed-rest. This may be undertaken at home under the care of the family doctor or more effectively in hospital; undoubtedly rest in hospital confers additional benefit, possibly because of the release from domestic and business worries. Whatever be the reason, pain usually disappears after a few days of bed-rest in hospital.

2. Diet

It is usually said that a suitable diet for ulcer patients should be mechanically and chemically non-irritating and should consist of small frequent meals. However, there is little experimental evidence to support this advice, it has been shown, for example, that hourly feedings of milk provoke more acid secretion in the stomach than does the ordinary routine off our meals a day, and there is no evidence that the rate of ulcer healing can be accelerated by the traditional bland ulcer diet. Indeed, persistence with such diets may be harmful since they may lead to suboptimal take of vitamin C. Never the less, when symptoms are severe, the patient often appreciates frequent feeds and is helped by dietetic advice.

3. Tobacco smoking

There is good evidence to show that suspension of smoking accelerates the healing of gastric ulcers, and it is likely that this also applies to duodenal ulcers. Thus, tobacco smoking should be prohibited and totally abandoned.

Drugs

Antacids

Give symptomatic relief. Recent studies show that they promote healing and reduce recurrences. Mainly there are two types of antacids.

(i) Absorbable antacids

Sod. bicarbonate and Cal bicarbonate for short term intermittent relief. Continuous use may cause alkalosis or milk alkalosis syndrome. Since symptoms of this complication are not distinctive (nausea, headache, weakness), the disorder may progress unrecognized causing kidney damage. These antacids should not be used by patients who vomited, who are dehydrated, or who have hypertension or renal function impairment. There is evidence that calcium salts increase acid secretion by local action on the gastric mucosa.

(ii) Non-absorbable antacids

(a) Al. Hydroxide is relatively safe and very commonly used. However there is slight risk of phosphate depletion. Symptoms include anorexia, weakness and malaise. If the loss is very severe, then long term use may result in osteomalacia.

(b) Magnesium salts may cause diarrhea. Some compounds contain either magnesium and aluminum hydroxide or al. hydroxide and magnesium trisilicate. Mg. preparation should be given with caution to patients with renal damage.

(c) Anticholinergics: Are given mainly to delay emptying of the stomach and thus prolonging antacid retention. Commonly used preparations are poleline 4 mg. glycopyrrolate 1 mg. propantheline 7 to 15 mg. or isopropamide 5 mg. usually given orally in form of tablets. Anticholinergics can cause dry mouth and blurred vision. Difficulties in urinating, enlarged prostate, evidence of acute narrow angle glaucoma are some of the common contraindications for anticholinergics.

(d) Histamine H2 receptor blocking agents: Most commonly used therapeutic agent. They act as a competitive inhibitor of histamine at H2 receptors. In the stomach it blocks gastric acid secretion stimulated by histamine, gastrin (a gastric hormone), parasympathetic activity and food and diminishes both basal and nocturnal gastric acid secretion. Pepsin, secretion, gastric juice volume and intrinsic factor secretion are also reduced. In the gallbladder, H2 blockers potentiate chokeystokinin induced contraction.

The first compound which was in common use was Cimetidine. Newer compounds like Ranitidine and Famotidine are also now in use.

With Cimetidine the symptoms are commonly relieved within the 1st week, healing may take 2 to 8 weeks. The dose for Cimetidine is 800 mg. q.i.d. (with meals and at bed time).

Though cimetidine is generally well tolerated, a mild increase in serum creatinine and serum transaminases often occurs with apparent clinical significance. Diarrhoea, rash, drug fever and myalgias have been reported and mental confusion with agitation may develop in elderly patients with renal impairment. Other rare side-effects include illness in patients with extensive burns, sinus bradycardia, hypotension after rapid I. V. injection and hyperglycemia.

(e) Secralfate: Is a sucrose and aluminum containing disaccharide. It is reported to combine with proteins and proteolytic enzymes (*e.g.*, pepsin) and to form in the base of the ulcer a protective coating that promotes healing.

(f) Omeprazole: Blocks gastric acid secretion by inhibiting H, K - ATP use in parietal cell membrane. This is used in the treatment of peptic ulcer.

Omeprazole produces nausea, diarrhoea, abdominal colon paraesthesia, dizziness. Toxicological studies indicated that the drug form carcinoid tumors in gastric mucosa in animals.

(g) Radiation therapy and surgery: For certain cases of duodenal ulcer.

PLANTS FOR ANTI-ULCER

1. *Adhatoda vasica* Nees. (Acanthaceae)

Common names

Hindi: Arusha
Beng.: Bakash or Vasaka
Guj.: Adulso

Distribution

It is sub-herbaceous bush, found throughout the year in plains and sub-Himalayan tracts in India, ascending up to

1200 metres, flowers during February-March and also at the end of rainy seasons.

Parts used: Leaves.

Pharmacological activities

Produced anti-ulcer activity in the form of herbal or herbomineral preparations.

Chemical constituents

See Anti-diabetic Chapter.

2. *Aloe barbadensis* Mill. (Liliaceae)

Common names
Eng.: Curacao Aloe, Barbadose Aloe
Hindi: Ghee-Kuvar
Berlg.: Ghrita-Kumari

Distribution

It is xerophylic, arborescent of herbaceous, the fleshy and strongly cuticularised leaves usually pricky at the margin, and arranged in dense rosettes. It is naturalized in India. It is planted in many Indian gardens and available all over India.

Parts used: Leaves.

Pharmacological activities

Oral administration of the plant extract significantly reduced both the number of ethanol induced gastric lesions as well as lesion index in experimental rats. It produced anti-ulcer activity in experimental rats.

Chemical constituents

See Anti-diabetic Chapter.

3. *Alpinia galanga* Wild. (Zingiberaceae)

Common names
Eng.: The greater Galangal
Hindi: Kulanjan
Beng.: Kulanjan

Distribution

Found in South India and Bengal.

Parts used: Rhizomes.

Pharmacological activities

Ethanolic extract exhibited gastric anti-secretory anti-ulcer and cytoprotective activities in rats.

Chemical constituents

See Anti-diabetic Chapter.

4. *Althaea rosea* Cav. (Malvaceae)

Common names
Eng.: Holly-Hock
Hindi: Gulkhairu

Distribution

A herb, often grown in gardens for showy flowers, which yield a red dye.

Parts used: Stems.

Pharmacological activities

The damages of mucus membrane of rat stomach, caused by aspirin, atophan considerably decreased by the polysaccharides obtained from stem.

Chemical constituents

3-Glucoside of kaempferol, cyanidin, kaempferol, quercetin, flavonolglycoside-herbacin and 3-rutinoside were isolated from the plant.

5. *Antirrhinum majus* Linn. (Scrophulariaceae)

Common name
Eng.: Snapdragon

Distribution

Cultivated as garden plant.

Parts used: Leaves

Pharmacological activities

Leaf extract showed marked anti-ulcer action when given orally to guinea pigs at 10 mg/kg/day during 5 days of histamine treatment to 8 rats for 12 days preceding pyloric ligation; under same conditions, apigenin was much less effective.

Chemical constituents

Alkaloid-4-methyl-2,6-naphthyridine three other tertiary alkaloids, sixteen amino acids such of which eleven were identified as alanine, aspartic acid, cysteine, glutarnic acids, glutamine, glycine, lysine, serine, threonine tyrosine and valine were isolated from aerial parts and cyanidin-3-glucoside and cyanidine-3- rutinoside from flowers.

6. *Asparagus racemosus* Willd. (Liliaceae)

Common names
Hindi: Satavar
Beng.: Satamuli

Distribution

Found throughout tropical and subtropical parts of India, up to 4000 ft in the Himalayas.

Parts used: Roots.

Pharmacological activities

Produced anti-ulcer activity in the form of herbal or herbo mineral preparation. Oral administration of powder of dry roots prevented the formation of duodenal ulcer in rats. Root exhibited ulcer healing effect in patients probably *via* strength-ending the mucosal resistance or cytoprotection. It did not produce antacid activity.

Chemical constituents

Leaves contain quercetin-3-glucuronide, of m. p. 204°.

7. *Azadirachta indica* A. Juss. Linn. (Meliaceae)

Common names
Eng.: Margosa Tree
Hindi: Neem
Beng.: Neem

Distribution

A large tree with rough bark. Native to India, grown all over India, grows wild in the dry forests of the Decan.

Parts Used

Leaves, Fruits and Seeds.

Pharmacological activities

Aqueous extract of leaves, exerted anti-ulcer effect in rats on oral administration. Aqueous extract of leaves produced anti-ulcer activity in rats exposed to cold restraint stress or given ethanol orally. A significant anti-ulcer effect was found in nimbidin, one of the constituents of oil of the seed. Some active ingredients isolated from the lipid parts of fruits, exhibited anti-ulcer activity in stress induced ulcers in male rats.

Chemical constituents

See Anti-inflammatory Chapter.

8. *Benincasa hispida* (Thunb.) Cogn. (Cucurbitaceae)

Common names
Eng.: White squash or Ash Gourd
Hindi: Petha
Beng.: Chalkumra

Distribution

A large climber and annual plant. Cultivated throughout India, fruits used as vegetable. Flowers are large yellow.

Parts used: Fruits.

Pharmacological activities

Fruits produced anti ulcer activity in shay rats.

Chemical constituents

See Anti-diabetic Chapter.

9. *Boerhaavia diffusa* Linn. (Nyctaginaceae)

Common names
Eng.: Spreading Hog-weed
Hindi: Sant
Beng.: Rakta-punarnava

Distribution

A perennial creeping weed, with pinkish flower found in almost all parts of India.

Parts used: Aerial parts.

Pharmacological activities

It showed protection against stress induced ulcer in albino rats/mice.

Chemical constituents

See Anti-inflammaztory Chapter.

10. *Brassica oleracea* Linn. Var. Capitata Linn. (Brassicaceae)

Common names
Eng.: Cabbage
Hindi: Band-Gobi
Beng.: Bandhakapi

Distribution

Cabbage is a commonly used as vegetable like cauliflower and available throughout India.

Parts used: Leaves.

Pharmacological activities

Leaf powder did not affect the ulcer index significantly, but its aqueous extract reduced the index.

Chemical constituents

See Anti-diabetic Chapter.

11. *Catha edulis* Forsk. (Celastraceae)

Common name
Eng. African tea.

Distribution

An evergreen shrub, native etc South Africa, introduced into India at Mysore and Bombay. Leaves and buds have stimulating effect.

Parts used: Aerial parts.

Pharmacological activities

Extract produced anti-ulcer activity and a non-specific neuro-muscular blocking effect.

Chemical constituents

Leaves and young shoots contain d-norpseudo ephedrine and amino acids, beside this, an alkaloid, ephderine and cathine (d-norpseudoephedrine) were isolated.

12. *Centella asiatica* (Linn.) Urban. (Apiaceae)

Common names
Eng.: Indian pennywort
Hindi: Brahmamanduki
Beng.: Tholkuri

Distribution

A herb found throughout India.

Parts used: Leaves.

Pharmacological activities

Plant extracts inhibited significant gastric ulceration induced by cold and restraint stress in rats.

Chemical constituents

Thankunic acid, m. p. 314°, triterpene glycoside than kuniside m.p. acetic acid as methyl ester, brahmic acid, m.p. 293, isobrahmic acid, m.p. 263°, brahmoside, m.p. 242°, and brahminoside, m. p. 223°, isothankuniside, m.p. 250°, isothankunic acid, m. p. 288°, made-cassoside, m. p. 220°, and made-cassic acid, m.p. 265°. A new triterpene acid-isolated and characterized as 2α, 6/3-trihydroxynes-12-enoic acid: polyacetylenes and nine other acetylenes isolated from subterranean parts.

13. *Cephalandra indica* N. (Cucurbitaceae)

Common names
Eng.: IvyGourd
Hindi: Kanduri
Beng.: Telakucha

Distribution

Found throughout India, wild and cultivated.

Parts used: Leaves.

Pharmacological activities

Aqueous, alcoholic extract and a crystalline compound of the leaf produced anti-ulcer effect in white albino rats.

Chemical constituents

See Anti-diabetic Chapter.

14. *Cinnamomum cassia* Blume. (Lauraceae)

Common names
Eng.: Casialignea

Distribution

An evergreen tree with aromatic bark. Bark is used as stomachic and carminative.

Parts used: Barks.

Pharmacological activities

Cassioside and cassiol, two components obtained from bark produced anti-ulcer activity.

Chemical constituents

Bark yields volatile oil known as 'Oil of cassia", contain high percentage of cinnamic aldehyde.

15. *Cocos nucifera* Linn. (Arecaceae)

Common names

Eng.: Coconut
Hindi: Nariyal
Beng.: Narikel, Dab

Distribution

The tree is cultivated mostly in hot and humid parts of India, particularly near sea, for its great commercial value.

Parts used: Seeds.

Pharmacological activities

Produced anti-ulcer activity in the form of herbal or herbo mineral preparations.

Chemical constituents

Coconut contains protamine, albumine, composed of glutamic acids, alanine, serine, cystine, leucine, isoleucine, valine, aspartic acid and other amino acids, globulin. Detection of phenol, p-cresol, caproic acid and p-hydroxybenzoic acid by TLC in shell fibres; in addition, tar from shells contained crotonaldehyde, furfural and acetic acid.

16. *Curcuma zedoaria* Rosc. (Zingiberaceae)

Common names

Eng.: Zedoary
Hindi & Beng.: Kachura

Distribution

A stem-less herb, root stock is tuberous. Rhizomes are considered as stimulant, carminative and stomachic. The herb is found in some parts of India.

Parts used: Rhizomes.

Pharmacological activities

A significant inhibition in ulcer formation was produced on oral arid subcutaneous administration of the extract in restrained and water immersed mice. Chemical constituents: Rhizome contains zedoarone, curcolone, curcumenol, a furanodiene, m. p. 440 and beside these sesquiterpenoid-zederone, pyrocurzerenone m. p. 76° and curcumol, m. p. 1410. Isolation of main component of essential oil-curzerenone- and its structure elucidation; absolute structure of zederone established; another sesquiterpene-dehydrocurdione isolated and characterized; synthesis of pyrocurzerenone; curzerenone, pyrocurzerenone and new furanosesquiterpenoids-furanodienone (I), isofuranodienone (II) and epicurzerenone-isolated and their absolute configurations established.

17. *Eclipta alba* Linn. (Asteraceae)

Common names

Hindi: Bhangra, babri
Beng.: Kesuti, keshukti

Distribution

Herbs found throughout India, particularly in moist ground. It is used for darkening hair.

Parts used: Aerial parts.

Pharmacological activities

Produced anti-ulcer activity in the form of herbal of herbo mineral preparations.

Chemical constituents

It contains demethyl wedelolactone-7-O-glucoside, sixteen polyacetylene thiophenes and nicotine.

18. ***Emblica officinalis*** **Gaertn. (Euphorbiaceae)**

Common names
Eng.: Emblic Myrobalan
Hindi: Amla, Amlika
Beng.: Dhatri, Amlaki

Distribution

A small or medium sized tree, found throughout tropical part of India, ascending up to 1300 m, cultivated in gardens and home yards.

Parts used: Fruits.

Pharmacological activities

Produced anti-ulcer activity in the form of herbal or herbomineral preparations.

Chemical constituents

See Anti-diabetic Chapter.

19. ***Ficus racemosa*** **Linn. (Moraceae)**

Common names
Eng. Cluster fig, country fig
Hindi: Gular
Beng.: Jagnadumur

Distribution

A large deciduous tree distributed throughout India particularly in evergreen forests, moist localities. It is cultivated in village for sauce and edible fruits.

Parts used: Barks.

Pharmacological activities

Anti-ulcer and anti-secretory activities were produced in experimental rats by the aqueous extract of the dried bark.

Chemical constituents

See Anti-diabetic Chapter.

20. ***Glycyrrhiza glabra*** **Linn. (Papilionaceae)**

Common names
Eng.: Liquorice
Hindi: Mulathi
Beng.: Jashtimadhu

Distribution

A perennial native to the Mediterranean region and is now grown as a herb, cultivated in south India. Punjab and Kashmir.

Parts used: Rhizome.

Pharmacological activities

Anti-ulcer effect was observed in it. Produced anti-ulcer activity in the form of herbal or herbo mineral preparations. Liquorices may be applied for symptomatic relief from the pain of peptic ulcer. Anti-ulcer activities along with other pharmacological properties of liquorices were described.

Chemical constituents

The chief constituent of liquorice is glycyrrhizin, which is present in the drug in the form of potassium and calcium salts of glycyrrhizic acid. It also contains glucose (3.8%), sucrose (2.4-6.5%), bitter principles, resins, asparagine (2-4%) and fat (0.8%). A flavanone rhamnoglucoside, m.p. 232°, isolated from roots; used as smooth muscle relaxant in pharmaceutical composition; twenty seven flavonoids present in roots of these sixisolated and three identified as 4-7-dihydroxyflavone (sitosterol); its 4-j3-D-glucoside (liquiritin) and 2,4,-trihydroxychalcone (isoliquiritigenin); other three are new flavonoids, -L-1, m. p. 164°, L-5, m. p. 150°, L-7, m. p. 142°, aglycones (90%) obtained by hydrolysis of flavonoid L-1, separated into liquiritigenin, m. p. 207° and isoliquiritigenin, m.p. 198°, two new chalcone glycosides characterised as trans-isoliquiritigenin4'- 3-D-glucopyranoside (isoliquiritin), m. p. 230°, liquiritigenin, liquiritin and isoliquiritigenin detected by PC: 7-hydroxy-4-methoxyisoflavone (formononetin) from roots: a new flavonoid glycoside-licuraside, m.p. 1 50° characterized as trans-isoliquiritigenin-4-O-(-O-Beta-D-glucopyranosyl-2—D-apiofuranoside); ten flavonoids identified in plant by PC and saponaretin (isovitexin) characterised all; isolation and structure of new flavone glycoside-rhamnoliquiritin. A new triterpenoid liquorice acid-isolated from roots, its structure elucidated: isolation of two triterpenoid acids-11-deoxoglycyrrhetic acid and liquiritic acid, m.p. 298°, characterized as C-20 epimer of glycyrrhetinic acid; structure of isoglabrolide, m.p. 318°: new lactones-glabrolide, m. p. 360°, deoxoglabrolide, m. p. 274° and isoglabrolide-obtained from acid hydrolysate of crude glucosides, glycyrrhizic acid isolated from rhizomes and roots; isolation of glycyrrhetol, m.p. 304° and 21 α-hydroxy isoglabrolide and their structures elucidated; synthesis of glycyrrhizic acid derivatives, 18-hydroxyglycyrrhetic acid isolated from acid fraction of plant extract; 24-hydroxy-1 1-deoxyglycyrrhetic and 24-hydroxyglycyrrhetic acids isolated as their Me esters; structures of 24-hydroxyliquiritic and liquiridiolic acids determined: preparation of Me glabrate from Me glycyrrhetate.

21. *Leucas aspera* Spreng. (Lamiaceae)

Common names

Hindi & Beng.: Chhotohalkusha

Distribution

Awooly herb, found throughout the plain part of India.

Parts used: Whole plant.

Pharmacological activities

Anti-ulcer activity was observed.

Chemical constituents

Two alkaloids rn. p. 139° and rn. p. 183° galactose, ursolic acid, -sitosterol, oleanolic acid and two sterols m.p. 350 and 13O, α-sitosterol, sitosterol and a compound A, m. p. 61° isolated from aerial parts.

22. *Mikania cordata* (Burrn.) B.L. Robinson

Common name
Bang.: Laralata

Distribution

A herbaceous climber, found in South India, Assam and West Bengal.

Parts used: Leaves.

Pharmacological activities

Potent antiulcer properties were found in rats by the plant extract.

Chemical constituents

Mikanolide m. p. 230° from leaves and stems, flavone-mikanin, (3,5-dihydroxy-6,7,4'--trimethoxyflavone), fumaric acid and epifriedelinol from roots, leaves and stems, and fridelin, and stigrnasterol from roots were isolated. Six new sesquiterpene dilactones-mikanolide, dihydromikanolide, deoxymikanolide, scandenolide, dihydroscandenolide and miscandenin isolated from aerial parts, crystal structure of mikanolide, three new labdenin acid derivatives, two kaurenoic acid derivatives and four new germacranolides isolated and their structures determined.

23. *Musa paradisiaca* Linn.(Musaceae)

Common names
Eng.: Edible Banana Plantain
Hindi: Kela
Beng.: Kala

Distribution

It is native in India, and cultivated for its fruits.

Parts used: Fruits

Pharmacological activities

Anti-ulcerogenic activity was produced in rats by the powder of plantain banana of unripe fruits on oral administration. Two sterylacylglucoside, active against peptic and duodenal ulcers were isolated from the fruits. Produced anti- ulcer effect in the form of herbal or herbomineral preparations.

Chemical constituents

The fruits are rich source of carbohydrates, mineral and vitamins (specially B-complex) I 4-Methyl-9 Beta, 19-cyclo- 5α -ergost-24(28)-en-α-3 Beta-ol isolated from plant.

24. *Ocimum basilicum* Linn. (Lamiaceae)

Common names
Eng.: Common basil
Hindi: Gulaltilsi

Distribution

Indigenous to the lower hills of Punjab, cultivated throughout the greater part of India. Ban-tulsi.

Parts used: Aerial parts.

Pharmacological activities

Its powder, aqueous extract and ethanolic extract reduced the ulcer index.

Chemical constituents

Its volatile oil contains Ocimene, methylchavicol, sambulene, methyl cinnamate, linalool, borneol, safrole and cineole. Sesquiterpene hydrocarbon-1-

epibicyclosesqui- phellandrene, methyl charriol (90%) and linalool were also isolated.

25. *Panax ginseng* Mey. C.A. (Araliaceae)

Common names
Eng.: Chinese ginseng, Asiatic ginseng

Distribution

This plant does not occur in India, but P. pseudo ginseng Wall. and few of its varieties are available at different parts of the Himalayas. The root of *P. ginseng* is a very important and valuable drug in Chinese system of medicine.

Parts used: Leaves.

Pharmacological activities

A polysaccharide obtained from leaves, produced anti-ulcer activity.

Chemical constituents

Root contains panaxadiol, m. p. 2500, a polysaccharide, two triterpene sapogenins, oleanolic acid, 13-sitosterol, three other glucosides and saponin-ginsenoside Rg-asdeca-acetate. Beside these the herb also contains flavonoids kaempferol and trifolin, panaxosides A and B, steroidal hormones, pantothenic acid, niacin and an alkaloid. Panaxosides B and C isolated from roots found to be trisaccharides, carbohydrate moiety of former contained glucose units whereas that of latter contained glucose and rhamnose two of these units were linked panaxatriol isolated; ginsenoside-Rgl isolated from roots and its structure established; ginsenosides Rbl, Rb2 and Rc, m.p. 1970, 200° and 199° respectively, on hydrolysis yielded 20(S)-protopanaxadiol, partial hydrolysis of these saponins yielded prosapogen in which was identified as 3-0-(2 Beta-D-glucopyranosyl-3-D-glucopyranosyl (20S)-protopanaxadiol; anti-inflammatory glucoside spanaxsaponins A and C isolated, partial structure of former proposed; preparation of a cardiotonic substance which contains eight triterpenoid saponins; ginsenosides RO, m.p. 239° (chikusetsusaponin V), Rbl Rb2, Re and Rd, m. p. 206° isolated from roots and their structures established; ginsenosides Re, Rf isolated along with panaxadiol, panaxatriol, daucosterol, mannitol, sucrose and glucose; ginsenosides Rd, Re, Rgl, isolated from flowers and buds; new saponins ginsenosides Fl, F2 and F3 isolated from leaves and their structures established; crude saponins isolated from aerial parts (leaves, 12.8, stems, 1.6 and flowers, 6.9%); crystal structure of panaxoside A; new saponins ginsenoside Rhi and M7cd isolated from roots and flower buds respectively and characterised as 6-0- Beta-D-glucopyranoside of dammar 25-ene-33-6c, 1 2b, 20(S), 24S-pentanol respectively; ginsenosides Rbl, Rb2 and Re isolated from leaves and flower buds; ginsenoside F3 from flower buds; ginsenosides Rb2, Re, Rd, Re and Rgl isolated from fruits; in addition to above, ginsenosides Ra and Rg2 isolated from roots.

26. *Plantago asiatica* Linn. P. Major Hook. F. (Plantaginaceae)

Common name
Kan.: Sirapotta gida

Distribution

Herbs- commonly found at hilly parts of India, on waste places.

Parts used: Leaves.

Pharmacological activities

A water-soluble substance, plantaglucide,

isolated from leaves, exhibited anti-ulcer property.

Chemical constituents

Plantaglucide.

27. *Pluchea indica* Less. (Asteraceae)

Common name

Beng.: Manighu rukha, Kukronda

Distribution

Shrubs, found in salty marshes in Sundarbans (W.B). Distributed in new world tropical and sub-tropical regions of Asia. It contains around 50 species out of which 6 have been recorded.

Parts used: Roots.

Pharmacological activities

Root extract exhibited significant antiulcer activity in rats.

Chemical constituents

See anti-inflammatory Chapter.

28. *Solanum nigrum* Linn. (Solanaceae)

Common names

Eng.: Black night shade
Hindi: Makoi
Beng.: Kakmachi

Distribution

A shrub, found throughout India, berries are globose and green, but turns yellow on maturity, flowers are white a good source of solasodine.

Parts used: Aerial parts.

Pharmacological activities

Powder and methanolic extract significantly lowered the ulcer index.

Chemical constituents

Solasodine like compounds in leaves and berries, tigogenin in berries, glucoalkaloids in immature berries, and solasonine and solamargine in leaves were detected.

29. *Strychnos ux vomica* Linn. (Loganiaceae)

Common names

Eng. Nux-vomica, snake-wood
Hindi: Kajra, Kuchla
Beng.:Kuchila.

Distribution

A tree found almost throughout the tropical parts of India. Dried and ripe seeds are used drug as tonic, stimulant etc.

Parts used: Seed.

Pharmacological activities

Oral administration of the powder produced anti-ulcer activity in shay rats. Strychnine exhibited anti- ulcer effect in shay rats.

Chemical constituents

Two major alkaloids of seeds are strychnine and brucine, leaves contain strychnine, brucine, methoxy-strychnine and vomi-cine, fruits contain loganin, m.p. 222°, a glycoside.

30. *Terminalia chebula* Retz. (Combretaceae)

Common names

Eng.: Chebulic myrobalan
Hindi: Harara, Harad

Beng.: Haritaki

Distribution

A large tree, found throughout India in deciduous forests. Fruits are of variable size, and rich in tannin.

Parts used: Fruits.

Pharmacological activities

Powder of dry fruits, on oral administration, prevented the formation of duodenal ulcer and reduced the ulcer index in rats, in experimentally induced acute gastric ulcerations.

Chemical constituents

Fruits contain terchebin a tannin compound and a glycoside of anthraquinone derivatives, and flowers contain chebulin, m. p. 249°. Palmitic, stearic, oleic, linoleic, arachidic and behenic acids isolated from fruit kernels.

31. *Trichosanthes dioica* Roxb. (Cucurbitaceae)

Common names
Eng.: Pointed gourd
Hindi: Parwal
Beng.: Potol

Distribution

An annual creeping herb, cultivated throughout the plains of N. India, extending to Assam and Bengal.

Parts used: Leaves.

Pharmacological activities

Produced anti-ulcer effect in the form of herbal or herbomineral preparations.

Chemical constituents

Roots contain an amorphous saponin, hentriacontane, or phytosterol, a nitrogenous bitter principle, glycosidic in nature and resembling colocynth, small amount of essential oil, little fixed oil, and traces of tannins.

32. *Trigonella foenum-graceum* Linn. (Fabaceae)

Common names
Eng.: Fennugreek
Hindi: Methi
Beng.:Methi

Distribution

Cultivated in many parts of India.

Parts used: Seeds.

Pharmacological activities

Anti-ulcer activity was evaluated in rats.

Chemical constituents

See Antifertility Chapter.

33. *Zingiber officinale* Rose. (Zingiberaceae)

Common names
Eng.: Ginger
Hindi: Adrak
Beng.: Ada

Distribution

The Plant is a herbaceous perennial, producing leafy shoots which attain a height of about 1-3 ft. ginger is cultivated in many places. Cochin ginger takes the highest rank among Indian gingers but the districts of

Rungpur, Midnapore and Hooghly in Bengal, Surat and Jhama in Bombay and Kumaon in Uttar Pradesh are also noted for production of good ginger.

Parts used: Rhizomes.

Pharmacological activities

On oral administration of acetone extract of the rhizome, zingiberone, the main terpenoid of acetone extract and gingerol, the pungent principle of the rhizome significantly prevented gastric lesions in HCI/ethanol induced gastric lesion in rats. Beta sesquiphellandrene, beta-bisabolene and 6-shogaol, some active principles of Taiwan zinger produced anti-ulcer activity in HCI/ ethanol induced gastric lesions in rats.

Chemical constituents

Detection of heptone, octane, isovaleraldehyde, nonanol, ethyl pyrene, camphene, 3-pinene, sabiriene, myrecene, limonene, 3-phellandrene and 1 ,8-cineole in essential oH by GLC; presence of gingediol, methylgingediol and their diacetates by GC-MS; new sesquiterpenes-sesquiterpene, cis-sesquisabinese hydrate and zingiberene (2 methyl-6(trans-4'-methyl-4' hydroxycyclohex-2-enyl)-hept-2-ene) isolated and their structures determined; car-3-ene) α-terpinene, α-terpineol, nerol, 1, 8-cineole, zingiberene, nerol geraniol, geraniol and geranyl acetate identified in essential oil from rhizomes.

4

MEDICINAL PLANTS FOR ANTIFERTILITY

INTRODUCTION

Fertility Control

The incredible growth of the world population stands as one of the significant events of the modern era. The current world population of about 5 billion is expected to be 6 billion by year 2000; it is worth while to mention that most of the growth will be in under developed countries. The old Testament dictum "Be fruitful, and multiply" has been religiously followed by readers and non readers of the Bible alike. In 1798, Thomas Robert Malthus started a great controversy by opposing the prevailing view of unlimited progress for man by making two postulates and a conclusion. He postulated that "food is necessary to the existence of man", and that sexual attraction between woman and man is necessary and likely to persist, since "towards the extinction of the passion between the sexes, no progress what ever has hitherto been made, "barring" individual exceptions. Malthus concluded that "the power of population is in equality greater than the power in the earth to produce subsistence for man," "a natural inequality "that would some day loom "insurmountable in the way to the perfectibility of society." Malthus' essay sparked great controversy and inquiry into the principle governing the growth of population. In seeking to discover the causes of population increase, T.R. Edmonds in 1832 suggested that "deterioration in the condition of the English labors...., the destruction of the feeling of self-respect" was of such a great distress that "among the great body of the people..., sexual intercourse is the only gratification.... When they are better fed they will have other enjoyments at command than sexual intercourse, and their numbers.... will not increase in the same proportion as at present." Today we realize that our sheer numbers have increased so much that they are straining Earth's capacity to supply food, energy, and raw material. We also know, perhaps better than T.R. Edmonds, where some of the blames for this growth lie. Advances in medicine and public health have led to a significant decline in mortality and an increased life expectancy. Thus, medical science has began to assume a portion of rate of the responsibility for population explosion.

Modern Methods to Control Fertility

A number of methods are available to control fertility. These can be classified into five categories.

A. *Barrier methods*

The most widely used barrier device is condom. Technological developments have evolved in extra thin membrane sheaths marketed in a variety of colors with or without extra lubricants. When used properly, condoms have a satisfactory efficacy; no medication is involved in either male or female. An extract advantage in the use of condom is the lack of overt transmission of venereal disease.

Diaphragm

Usable in the female, these are sacks to cover the passage of the sperm to the cervix. It has a fairly high failure rate presumably because the sizes vary and optional fitness is not achieved. Its use is on the wave.

Along with the diaphragm, the use of foams, jellies and creams containing spermicides is invariably recommended to increase the overall efficacy.

B. *Intra uterine devices (IUD)*

The anti-fertility action of a foreign body in the uterus was known since ancient times in Egypt and India. A small stone was inserted in the uterus of caravan camel toward off pregnancy during long journey, IUD were introduced in the family planning programme about 25 years back. A variety of IUD's differing in shape has been proposed. Most of them are made of silastic polymers with or without additional metallic components. Examples are Lippe's coop, the Dalkon shield, Soonawala Device etc.

The first generation IUD had limited acceptors inspite of advantage of one time insertion for a long period of protection. The bleeding disorder and pain are known to occur. These side effects have been reduced in the 2nd generation of IUD currently offered, which are Cu-IUD and the copper-Silver IUDs. The protective efficacy is also better in comparison to the earlier ones.

Mechanism

It primarily prevents the implantation of the blastocyst. The device induces an inflammatory reaction in the endometrium. The Cu-IUD also has an additional action where they slow the release of Cu, influencing metabolic reaction of endometrial reaction of endometrial cells. Sliver in turn may give rise to galvanic currents.

C. *Contraceptive steroids (Oral contraceptives)*

Existence of a feedback control between the steroid hormones produced by ovaries and the pituitary gonadotropins which stimulate their synthesis, give rise to the property which formthe basis of utilizing progestogens and estrogens to inhibit the secretion of gonadotropins and thereby block the ovulations. Natural hormones, progesterone and estradiol were not employed since they have a short biological half-life. Instead of these, synthetic steroids with progestogenic and estrogenic activity were used. Earlier these drugs were administered by the oral rout, the oral pills consisting of a combination of a progestogens with small quantities of estrogens, taken daily for 21 days. However the incidences of side effects are high. A variety of major and minor side effects have been attributed to the use of oral contraceptives. The most concerned are cardiovascular side effects and induction of promotion of tumors. The risk of cardiovascular disorders increases several fold in women who the smokers.

In recent times injectable contraceptive steroids have been evolved.

These are depot preparation dispersed in an oily base.

The steroids commonly used are:

1. Medroxyprogesterone acetate or Depo-Provera
2. Norethindrone acetate

These two have 2 to 3 months span.

D. *Termination of Pregnancy*

E. *Sterilization, male/female*

New developments and improvements of the presently available methods

A. The Subdermal Implants Norplant

The population council has developed this implant, which consists of a set of six silastic capsules filled with the progestogen levonorgestrel. These are introduced beneath the skin under local anesthesia. The release of the steroid is at a steady rate and ensures effective contraception for a longer period but want to retain the option for another pregnancy. A simpler version of the same consists of two rods instead of six capsules may be available in near future.

B. Progestin Releasing IUD

New methods in trial stage

A. Analogues & Antagonist of Luteinizing Hormone Release Hormone (LHRH):

Several synthetic analogous of LHRH have prolonged biological half-life. Repeated intakes of these superagonists have the paradoxical action of desensitization of the target tissue receptors resulting in negation of the hormone effect. Daily administration of the agonists by systemic injections or nasal spray results in blockade of ovulation in the females, accompanied by diminution of ovarian progesterone secretion and amenorrhoea, since estrogen secretion however remains. These preparations may be contraindicated if it is extended for a long term. So, the treatment may be limited to one or two years, a spacing or gap procedure.

B. *Gossypol*:

This compound extracted from cotton seeds was used by accident in China by several thousand males, thereby causing azoospermia. The compound binds with sperm proteins. Its toxicity is high in rodents. However, in humans (as reported in China) only hypokalemia was found. Due to low efficacy versus toxicity ratio in conventional experimental animals, it is unlikely to be approved for human usage.

C. Prostaglandins and Antiprogesterone compounds for Menstrual Regulation and Termination of Pregnancy:

These contraceptive vaccines are in the developmental stage. This may merge as the most advanced birth control vaccine, having activity on HCG.

Herbal Options to Control Fertility

One of the serious problems of the third world particularly countries like India, is its geometrical increase in human population. According to estimation of WHO the population in India will exceed one hundred million at the end of this century. This population explosion will have negative impact on our economic policies and would simultaneously misbalance our socio-economic infrastructures. These would be manifested by unemployment, scarcity of food and shelter and also problem of soil, water and air pollutions would take its own toll in the long run. To combat the problem, initiations have been made by Government and non-Government institutions but progress of the object is less far reaching than from its target. In this regard, search for harmless and inexpensive oral agents for fertility control in human beings to reduce the population size is appreciable. In this context it will be appropriate to locate the large number of indigenous plants that are used as oral contraceptives especially by tribal and other sections. Such plants are even recommended in folk medicines and Ayurvedic medicines from very ancient times. Already several scientific papers have been published related to fertility control but still more plants are left. In the present context, an attempt has been made to gather information on the geographical distribution or such plants related to fertility control, their parts used, nature of chemical compounds of these plants and their pharmacological effects. Such information will cater industries in laying out projects on commercial basis. This would also help in searching plants in respective localities and know the chemical compounds with their definite identify in relation to fertility control.

PLANTS FOR ANTIFERTILITY

1. *Abroma augusta* Linn. (Sterculiaceae)

Common names
Eng.: Cotton Abroma
Hindi & Beng.: Olatkambal

Distribution

A large shrub or small tree, grown in gardens and widely distributed throughout hotter parts of India.

Parts used: Roots.

Pharmacological activities

Estrogenic activity in female albino rats abortification and antimplantation activity in mice. Uterotonic activity on isolated rat uterus on human and dog uterus *in situ*.

Chemical constituents

See Antidiabetic Chapter.

2. *Abrus precatorius* Linn. (Papilionaceae)

Common names
Eng.: Indian liquorice
Hindi: Chirmiti, Gumchi
Beng.: Chunhati

Distribution

Occurring throughout the country.

Parts used: Seeds.

Pharmacological activities

Seeds have antifertility activity in albino rats and Swiss mice; oxytocic activity *in vitro* in guineapigs, petroleum ether extract has 60% activity on late pregnancy in rats, antifertility effects on mice and rats. Indigenous preparations of seeds adversely influence pregnancy, virginal hemorrhage. Seeds oil showed post-coital anti-fertility activity in rats and oral contraceptives in mice and rats.

Chemical constituents

The roots and leaves contain glycyrrhizin (1.25%). Seeds contain an alkaloid abrine, $C_{12}H_{14}O_{21}N_2$, a glycoside abralin, $C_{13}H_{14}O_4$ and small quantity of fatty oil Abrine hypaphorine, choline, trigonelline, precatorine and methyl ester of N, N-dimethyl tryptophan methocation isolated from seeds; alcoholic extract of seeds showed presence of carbohydrates and amino acids whereas ammonium oxalate extract showed presence of sugars; 5B-cholanic acid isolated from seeds; two toxic antitumor proteins-abrin A and B isolated from seeds; three new isoflavan quinones I, II, and III isolated from roots and characterized.

3. *Achyranthes aspera* Linn. (Amaranthaceae)

Common names
Hindi: Puthkunda, chirchitta
Beng.: Apang
Sans.: Apamargah

Distribution

Occurring throughout India up to 3000 ft as a weed. It is also available in Balochistan.

Parts used: Whole plant, stem bark.

Pharmacological activities

Stem-bark abortification activity in mice.

Whole plant

Benzene extract of whole plant showed 100% abortification activity in rabbits.

Chemical constituents

Betaine (m.p. 292^0) was isolated from the plant. Ecdysterone (Polypodine A) from roots; two oleanolic acid based saponins from fruits and ecdysone from roots.

4. *Actiniopteris radiata* Linn. (Actinopteridaceae)

Common names
Eng.: Wild fern.

Distribution

Confined to tropical Africa.

Parts Used: Whole plant.

Pharmacological activities

Antifertility property in women (90%) given along with *Ocimum americanum* seeds.

Chemical constituents

Hentriacontane, hentriacontanol, Beta sitosterol, its palmitate and its glucoside, an unidentified glucoside, glucose and fructose from aerial parts.

5. *Adhatoda vasica* Nees. (Acanthaceae)

Common names
Eng.: Malabar nut
Hindi: Basak
Beng.: Basak

Distribution

Commonly found throughout India, especially in plants of W. Bengal.

Parts used: Leaves.

Pharmacological activities

Treatment with alkaloid vasicine showed abortification activity in guineapig, Uterotonic activity on human myometrium strip, abortificient activity in in guineapig. Leaves have no anti-fertility activity in mice and rats.

Chemical constituents

See Antidiabetic Chapter.

6. *Aeschynomene indica* Linn. (Papilionaceae)

Common names
Hindi: Laugauni
Beng.: Kath shola.

Distribution

A shrub occurring in canal banks, distributed in Kashmir, Bengal, Assam and throughout plains of India, ascending to 1500 m in hills.

Parts used: Whole plants.

Pharmacological activities

Spermicidal activity in human and rat semen, Saponin from whole plant shows spermicidal activity in human semen.

Chemical constituents

Vicenin-2, reynoutrin, rutin, myricitrin and robinin were isolated from plant.

7. *Albizialebbeck* L. (Mimosaceae)

Common names
Eng.: kokko, Lebbeck tree
Hindi: Sirls
Beng.: Sirish

Distribution

A common road side tree common in W. Bengal, Assam, Tamilnadu, M.P. Punjab and U.P.

Parts used: Seeds, Roots, Pods.

Pharmacological activities

Seed extracts shows antiovulatory activity of saponins when employed in rabbits. Roots without any anti-fertility activity. Pods and root extracts showed spermicidal activity in human and rat semen also saponin from pods and roots responded spermicidal activity in human semen.

Chemical Constituents

Echinocystic acid and B-sitosterol identified in bark and seeds. A new saponin-lebbekanin C on acid hydrolysis yielded echinocystic acid, glucose and rhamnose. Different saponins (lebbekanin_ A, m.p. 105^0, lebbekanin-D and lebbekanin-E m.p. 125^0), were isolated.

8. *Allium cepa* Linn. (Liliaceae)

Common names
Eng.: Onion
Hindi & Beng.: Piyaj

Distribution

A biennial herb commonly cultivated all over the country. Important onion growing states are Maharashtra, Tamilnadu, A.P., Karnataka and M.P.

Parts used: Bulb

Pharmacological activities

Bud extracts showed ecobolic effect in mice and rats.

Chemical constituents

See Antidiabetic chapter.

9. *Allium sativum* Linn (Liliaceae)

Common names
Eng. Garlic
Hindi: Lahsan
Beng.: Rasun.

Distribution

Native of central asia.

Parts used: Bud.

Pharmacological activities

Bulb extracts responded ecobolic in mice and rats; bud extract in other experiment shows oestrogenic activity in female albino rats.

Chemical constituents

See Anti-inflammatory Chapter.

10. ***Aloe barbadensis*** **Mill. (Liliaceae)**

Syn. ***A. vera*** **(L)** Burm. F

Common names
Eng.: Barbados aloe
Hindi: Ghikanvar, guar-patha
Beng.: Ghritakumari

Distribution

A stoloniferous secculent shrub, native to West Indies, now naturalized in India, also grown as ornamental in W. Bengal, Bihar and some other states. A salt resistant species is useful for side landscaping.

Parts used: Leaves, roots.

Pharmacological activities

Leaf extracts responded anti-fertility activity in mice and rats, also showed oestrogenic activity in female albino rats, aqueous extract shows anti-implantation and anti-ovulatory activity in rats and rabbits. Leaves and roots show anti-fertility activity in albino mice and rats. In another study, chloroform extract of the leaves responded significant reduction in fertility in female rats.

Chemical constituents

See antidiabetic chapter.

11. ***Aloe indica*** **Royle. (Liliaceae)**

Common name
Eng.: Aloe

Distribution

Plains of Bengal, a herb with sword shaped leaves.

Parts used: Leaves.

Pharmacological activities

Leaf extract shows oestrogenic activity in female albino rats. A significant increase in the fertility rate of experimental rabbits receiving 60 mg/kg more for aloe.

Chemical Constituents

Aloin is the principle active constituent of aloes which is a mixture of glycoside. Barblin is the main glycoside in aloin which is water soluble, besides this, aloin contains isobarbaloin, Beta barbaloin, aloe-emodin resins etc. The percentage of barbaloin in Indian aloes extracted from Aloe vera. Leafs shows that Indian species contained less quantity (3.8%) as compared to curacao aloe which contain 22.1% (27).

12. ***Anagallis arvensis*** **Linn. (Primulaceae)**

Common names
Hindi: Jonkmari
Guj.: Anagalide.

Distribution

An annual herb, bearing small red or white flowers occur in plains of India.

Parts used: Whole plant.

Pharmacological activities

Whole plant shows uterine stimulant activity on isolated uterus of guineapig, uterotonic activity on isolated uterine strips of rat, human and dog uterus *in situ*.

Chemical constituents

An acetyl saponin isolated from the plant. Isolation and identification of cucurbitacins B, D,E,I,L and R cucurbitacins gluscosides-arvenin I,II,II and IV-isolated and their structures' were established.

13. *Ananas comosus* Linn. (Bromeliaceae)

Common names
Eng.: Pine apple
Hindi: Ananas
Mal.: Kazhudhachakka.

Distribution

A biennial herb, native to S. America, now cultivated mostly in Tamilnadu, coastal Andhra Pradesh, Assam, Kerala, Karnataka, W. Bengal, Tripura and Orissa.

Parts used: Unripe fruits, leaves, rhizomes.

Pharmacological activities

Unripe fruits showing weak activity in pregnancy tests of rats, also anti-implantation activity in albino rats, rhizomes responded anti-fertility activity in albino rats and mice, roots have anti-fertility activity in mice rats, estrogenic activity in rats have been observed with wax from wastes.

Chemical constituents

Ripe fruits contain vitamin C (63 mg/100g). Ergosterol peroxide, 5-stigmastene-3B, 7a-diol isolated from leaves besides B-sitosterol, campesterol, stigmastanol and campestanol, Ergosterol peroxide, 5-stigmastene-3B,7a-diol isolated from leaves besides B-sitosterol, campesterol, stigmastanol and campestanol; 5-hydroxy tryptamine from leaves of crown of pineapple fruit, method for isolation of biologically active (anti-inflammatory, haemolytic) peptides from roots and juice.

14. *Anisomeles malabarica* R.Br. (Lamiaceae)

Common names
Hindi: Kala bhangra
Eng.: Malabar catmint.

Distribution

A woody herb or under shrubs found throughout India.

Parts used: Plant without root.

Pharmacological activities

Plants excluding roots show spermicidal activity in human and rat semen, saponin from plant (without root) responded spermicidal in human semen.

Chemical constituents

Ovatodiolide and anisomelic acid (anisomelolide) were isolated. New macrocyclic diterpeness-malabaric acid, 2-acetoxymalabaric acid anisomelyl acetate and anisomellol along with anisomelolide and ovatodiolide isolated and their structures were determined.

15. *Ardisia neriifolia* Wall. (Myrsinaceae)

Common names
Hindi: Kadna Banjam

Distribution

Ornamental shrub found in N. India and some other places in plains of India.

Parts used: Plants excluding roots.

Pharmacological activities

Plant extracts excluding roots showed spermicidal activity in human and rat semen.

16. *Areca catechu* Linn. (Arecaceae)

Common names
Eng.: Betelnut palm, Arecanut
Hindi: Supari
Beng.: Supari

Distribution

An erect palm native of Malaysia now grown along the coasts of Karnataka, Kerala, Tamilnadu, W. Bengal, Assam and Maharashtra.

Parts used: Nuts.

Pharmacological activities

Petroleum ether extract, alcoholic and aqueous extract show anti-implantation activity in albino rats, extract also shows abortificient activity in albino rats, Nut-oil also shows antifertility activity in female albino rats.

Chemical constituents

The nut contains arecain (0.1%), arecoline (0.07–0.1%), arecaieline, guvacoline, guvacine and choline occur only Nitrogenous substances "Avenacines" A and B (saponin) are obtained from leaves, stems and flowers.

17. *Aristolochia indica* Linn. (Aristolochiaceae)

Common names
Eng.: Indian birthwort
Hindi: Iswarmul
Beng.: Iswarmul

Distribution

A shrub found throughout the country mainly in the plains and lower hilly regions.

Parts used: Roots.

Pharmacological activities

Aristolic acid isolated from roots shows anti-implantation activity in mice.

Oral administration of p-coumaric acid isolated from roots at 50 mg/kg dose level produced 100 percent interceptive activity in mice.

Methyl ester of aristolic acid extracts from roots show 100% abortifacient activity in female mice, also roots extracts containing Aristolic acid showed marked antifertility activity in female albino rats. A sesquiterpene isolated from the roots showed anti implantation and anti-oestrogenic activities in female mice.

Chemical constituents

A sesquiterpene, Ishwarene, a sesquiterpene ketone Ishwerone and sesqiterpene alcohol Ishwarol have been isolated from the plant. The root contains a bitter yellow compound named as isoaristolochic acid ($C_{17}H_{11}O_7N$). An alkaloid aristolochene ($C_{17}H_{19}O_3N$)

crystalline powder, m.p. 215^0 C was isolated from the plants.

18. *Artabotrys odoratissimus* R.Br. (Annonaceae)

Common names
Eng.: Climbing ylang-ylang
Hindi: Hari champa
Beng.: Kath champa

Distribution

A large, scandent shrub with greenish-yellow, fragrant flowers often grows in gardens.

Parts used: Leaves.

Pharmacological activities

Leaf extracts show anti-fertility in albino rats. The drug is sage its effect is of long duration.

The drug also showed anti-estrogenic activity in rats, significant anti-implantation activity in female albino rats.

19. *Azadirachta indica* A. Juss. (Meliaceae)

Common names
Eng.: Margosa tree
Hindi: Neem
Beng.: Nim

Distribution

A common tree in the plains of W. Bengal and other regions in the plains of India.

Pharmacological activities

Aqueous extracts of bark causes immobilization of human and bovine spermatozoa. Application of sodium nimbinate from the seed oil shows strong spermicidal activity in rats and humans.

Chemical constituents

See Antidiabetic chapter.

20. *Bombax malabaricum* DC. (Bombacaceae)

Common names
Eng.: Red silk cotton
Hindi: Simul Semur
Beng.: Simul

Distribution

A tall tree, one of the finest trees in India, distributed throughout the country, most abundant in Assam, Andaman and W. Bengal.

Pharmacological activities

Uterotonic activity on isolated uterus of rats, guineapig and rabbit, dog and human.

Chemical constituents

Hydrolysis of gum yielded arabinose, galactose, galacturonic acid and traces of rhamnose. Partial hydrolysis gave 6-O-(B-D-galactopyranosyl-uronic acid)- D-galactopyranose. 2,3,5-tri and 2,5-di-O-methyl-L-arabinose identified as hydrolytic products of methylated gum.

21. *Bridelia retusa* Spreng. (Euphorbiaceae)

Common names
Hindi: Khaja
Beng.: Geio, Sans, Asana.

Distribution

Distributed throughout India.

Parts used: Bark.

Pharmacological activities

A mixture is made with pounded bark of the tree with gum of *Sterculia urens* Roxb. prescribed orally for 2-3 days after menstruation for complete infertility.

Chemical constituents

Bark contains tannin (16-40%).

22. *Butea monosperma* Lam. (Papilionaceae)

Common names
Eng.: Flame of the forest
Hindi: Dhak
Beng.: Palas.

Distribution

A small deciduous tree occurring throughout India, bears bright orange-red flowers.

Parts used

Leaves, flowers and seeds.

Pharmacological activities

Leaves responded chronic toxicity in dogs, rats and rabbits, pet. ether extract shows negative anti-fertility activity in rats, anti-implantation activity in mice, anti-implantation activity in rats, flowers extract in another experiment shows no anti-implantation activity in albino rats and in female rats, insignificant estrogenic activity in female albino rats.

Alcoholic extract of seeds responded significant anti-fertility activity in female rats but pet ether extract show no significant anti-fertility activity in female rats. Alcoholic extract of seeds show anti-implantation activity in mice, higher doses responded toxicity.

Chemical constituents

Seeds contain 18% of a fixed oil called moodooga oil, small quantity of resin and large quantity of water soluble albuminoids. Two new glycosides-monospermoside and isomono spermoside isolated together with butrin, isobutrin, coreopsis, isocoreopsin and sulfurein.

23. *Caesalpinia bonducella* Flem. (Caesalpiniaceae)

Common names
Eng.: Molucca bean
Hindi: karanju, Naktamala
Beng.: Nata karanja.

Distribution

A shrub distributed throughout the hotter parts of India, often grown as hedge plant.

Pharmacological activities

Seeds have anti-estrogenic activity in mice rabbits and anti-fertility action in mice and rats was found.

Chemical Constituents

Seeds contain, besides starchy, matter 25.13% of an oil, 1.925% of a bitter principle, 6.83% sugar, 3.791% salt, a-Caesalpin, m.p. 187^0, b-caesalpin, m.p. 243^0, g-caesalpin and d-caesalpin were isolated from seeds.

24. *Calamintha umbrosa* Fisch. & Mey. (Lamiaceae)

Common names
Sans.: Karidorna
Mal.: Karimthuma.

Parts used: Whole plant.

Pharmacological activities

Extracts of whole plant responded spermicidal activity in rat semen.

Chemical constituents

Leaves yield an essential oil.

25. *Calendula officinalis* Linn. (Asteraceae)

Common names
Eng.: Pot-marigold
Hindi: Zergul.

Distribution

An ornamental with bright orange yellow coloured heads.

Parts used: Whole plant.

Pharmacological activities

Extract of whole plant showed spermicidal activity in rat semen.

Parts used: Barks, leaves

Chemical constituents

Beta carotene lycopene, violaxanthin, rubixanthin, hentriacontane and two phytosterol were isolated. Calenduloside A, m.p. 260^0 isolated and identified as galactose-glycoside of oleanic acid. Ceryl alcohol stigmasterol, faradiol, m.p. 196^0 and calendar, m.p. 153^0 were isolated.

26. *Caltha palustris* Linn. (Ranunculaceae)

Common names
Punjab: Mumiri

Distribution

Punjab and other parts of India.

Parts used: Whole plant.

Pharmacological activities

Extract of whole plant shows spermicidal activity in human and rat semen, saponin from whole plant show spermicidal activity in human semen.

Chemical constituents

Saponin, heueborin and veratrin.

27. *Canscora constituents* Roxb. (Gentianaceae)

Common names
Hindi: Sankhapuspi
Beng.: Dankeni

Distribution

A herb occurring throughout India.

Parts used: Whole plant.

Pharmacological activities

Extracts fraction I & II showed spermicidal activity in human Semen and rat sperm suspension.

Chemical constituents

Compounds I and II were isolated from whole plants. Gluanone, conscoradione, friedelin, friedelim, friedelan-3B-ol, B-amyrin, sitosterol, stigmasterol and campesterol isolated from aerial parts; new xanthones-1,3,6,7-tetrahydroxyxanthone (l), 1,3,5,6-tetrahydroxy-3,5,6-trimethoxy and 1,6,7-trihydroxy-3-methoxyxanthose (lll) isolated and characterised; new xanthones-xanthone A, xanthone B and xanthone C isolated and characterized; structures of three xanthones revised to 1-hydroxy-3,5,6,7-tetramethoxy, 1,7-dihydroxy-3,5,6-trimethoxy and 1,6,7-trihydroxy-3,6-dimethoxyxanthone isolated from aerial plants; two new xanthone glycosides and two free xanthones isolated and identified as 1-glycosyloxy-3-hydroxy-5-methoxy-(VI), 7-glucosyloxy-1,6-dihydroxy-3,5-dimethoxy-(VII), 1,5,6-trihydroxy-3,7-dimethoxy-(VIII) and 1,5,7-trihydroxy-3,6-dimethoxy-xanthones (IX); structures of the two previously reported xanthone-4 and xanthone-12 redesigned as 1,3,5-trihydroxy-(X) and 1,5-dihydroxy-3,6,7-trimethoxy xanthones (XI) respectively.

28. *Capsella bursa*-Pastoris Linn. (Brassicaceae)

Common names
Eng.: Shepherd's Purse.

Distribution

A common weed of W. Bengal, also in some other regions of India.

Parts used

Leaves, dry powdered plant.

Pharmacological activities

Dry powdered plant impeded ovulation, developed temporary infertility in adult female and male mice when fed at 40% level in diet.

Chemical constituents

Contains an alkaloid bursin. The seeds yield fatty oil (35%). A new flavonoid-luteolin-7-rutinoside, m.p. 184^0, luteolin-7-galactoside, m.p. 228^0 and quercetin-3-rutinoside were isolated from bark.

29. *Carica papaya* Linn. (Caricaceae)

Common names
Hindi: Papita
Beng.: Pepe

Distribution

A small tree native to the West Indies and C. America cultivated chiefly in Assam, Bihar, U.P. Gujart and W. Bengal.

Parts used

Latex of green fruit, unripe fruit pulp, oil from pulp of unripe fruit, seeds.

Pharmacological activities

Petroleum ether and alcoholic extract of unripe fruit pulp

Responded 60% anti-implantation activity in albino rats, alcoholic and aqueous extracts are abortifacient in albino rats, also pet, ether extract showed significant anti-fertility activity in female albino rats, the oil from the pulp of unripe fruit responded anti-fertility activity in female albino rats, seeds extract also showed a decrease in fertility in albino mice but found to be highly toxic.

Chemical constituents

Analysis of the fruit gave moisture 89.6%, proteins 0.5%, and carbohydrate 9.5%. The fresh fruit pulp contains sucrose, invert sugar, a resinous substance, papain, malic acid and salts of tartaric and citric acids 1.2%. identification of carotenoids phytoene, phytofluene, B-carotene, cis-B-carotene, pigment X, 5,6-monoepony-B-carotene were isolated.

30. *Cedrus deodara* Roxb. (Pinaceae)

Common names
Eng.: Himalayan cedar, Deodar
Hindi: Deodar
Beng.: Devdareo

Distribution

A tall evergreen tree distributed in N-W. Himalayan from Kashmir to Garhwal. Forest of deodar occurs in Kulu, Kashmir, Chamba, Tehri-Garhwal, Almora, Shimla, Chakrata and Mussoorie hill station.

Parts used: Stems.

Pharmacological activities

Stem shows anti-fertility activity in female rats.

Chemical constituents

See Anti-inflammatory Chapter.

31. *Celsia coromandeliana* Vahl. (Scrophulariaceae)

Common names
Hindi: Gadartambaku
Beng.: Kukshima

Distribution

A weed commonly found in W. Bengal.

Parts used: Whole plant.

Pharmacological activity

Spermicidal activity in rat semen has been noted.

Chemical constituents

A new sterol-celsianol, m.p. 1660 isolated and its structure established as stigmasta 5,9-dien-3Beta-ol; a constant melting sterol, m.p. 1640 (celsianol) shown to be mixture o 5, 6 hydrostigma sterol and α-spinasterol; three new saponins celsiosides A, B and C isolated which yielded on hydrolysis celsiogenins A , B and C respectively; the sugars obtained in each case were glucose, fucose and arabinose inmolar ration 1:1:1, 1:1:1 and 2:1:1 respectively; structure of celsiogenin C determined as olean-11, 13-dien-3Beta, 22Beta, 23, 28-tetrol, were as oelsiogenins A and B identified as olean-12, 17-dien-3Beta, 23-diol and olean-11, 13-dien-3Beta 23, 28-triol respectively.

32. *Centratherum anthelminticum* L. (Asteraceae)

Common names
Hindi: Somraj
Beng.: Somraj, Kalijiri.

Distribution

A herb distributed throughout India.

Parts used: Seeds.

Pharmacological activities

Spermicidal activity in rat semen has been noted.

Chemical constituents

Contain a fixed oil, 7(Z)24-stigmastadienol, stigmasterol, 5-stigmasten-3Betal-ol and 7,22-stigmastadienol isolated from seeds; 8,14, (Z)24-stigmasterol acetate isolated from seeds; a new clemanolide-vernodalol, m.p. 133^0 isolated from seeds; amino acid composition of seeds determined.

33. ***Cicer arietinum*** **Linn. (Papilionaceae)**

Common names
Eng.: Gram, chick-pea
Hindi: Chana
Beng.: Chola.

Distribution

A much branched herb cultivated in U.P. Punjab, Haryana, Rajasthan, Bihar, M.P. as pulse crop.

Pharmacological activities

Seeds oil of gram seeds showed some oestrogenic activity in albino rats also oestrogenic activity of seeds in albino rats and seeds responded ecbolic property in mice and rats.

Chemical constituents

Contain a higher percentage of oil (4-5%). Isoliquiritigenin-4-Glycoside, 3'4'.7-trihydroxy flavone, daidzein, pratensein, p-coumaric acid, garbenzol and biochanin 7-glucoside from seedings. Biochanin A content implant 8.0 mg/kg; formononetin also detected; vanilic and p-hydroxybenzoic acids from leaves.

34. ***Cichorium intybus*** **Linn. (Asteraceae)**

Common names
Eng.: Chicory, Wild endive
Hindi: Kasni, kasani

Distribution

North-West part of India. The chicory is native of Europe found in India, cultivated else where.

Parts used: Whole plant.

Pharmacological activities

Restorative activity (84%) has been noted in female albino rats with plant extract.

Chemical constituents

See Antidiabetic Chapter.

35. ***Clerodendrum serratum*** **Linn. (Verbenaceae)**

Common name
Hindi: Barangi

Distribution

Sub Himalayan tract and ranges, ascending to 5.000 ft. Assam, W. Bengal, Singhbhum valley.

Parts used

Except root, whole plant and saponin from plant excluding root.

Pharmacological activities

Plants excluding roots showed spermicidal activity in human and rat semen also spermicidal activity in human semen has been observed with saponin from the plant excluding roots.

Chemical constituents

Glucose and D-(-)mannitol from root bark hydrolysis of crude saponin from bark yielded oleanolic acid, queretaroic acid and new serratagenic acid.

36. ***Conium maculatum*** **Linn. (Apiaceae)**

Common name
Eng.: Poison hemlock.

Distribution

South Africa, very poisonous biennial, stems dotted red.

Parts used: Whole plant.

Pharmacological activities

Tincture of the plant controls and inhibits the estrous cycle

In white albino female rats.

Chemical constituents

The plant contains alkaloid coniine.

37. ***Corchorus olitorius*** **Linn. (Tiliaceae)**

Common names
Eng.: Jute
Hindi: Pat
Beng.: Mithapat, desipat.

Distribution

A shrub grown in W. Bengal, Bihar, Assam and U.P.

Parts used: Seeds.

Pharmacological activities

In mice and rats seeds responded ecbolic property.

Chemical constituents

Structure of olitoriside from seeds elucidated as strophanthidin-3-B-D-boirinodido-B-D-glucoside, a new cardiac glycoside-corchoroside A from seeds; contained aglycones-strophanthidin and corchorgenin; glycosides-corchsularin, olitoroside and corchorosides A and B; corchoralic acid, B-sitosterol and a saponin; glycosides determined calorimetrically and main components of glycoside mixture olitoroside A separated by TLC; veticoside, m.p. 168^0 from seeds isolated.

38. ***Costus speciosus*** **Koening (Zingiberaceae)**

Common names
Hindi: Keu
Beng.: Keu
Sans.: Kemuka

Distribution

An ornamental herb widely distributed in Assam, N. Bengal, Khasi, Jaintia hills, sub-Himalayan parts of U.P. and H.P.

Parts used: Rhizomes.

Pharmacological activity

Administration of rhizome to albino rats showed estrogenic activity and also antifertility activity due to abortifacient properties in albino rats. Also have shown strong ecbolic property in rats, guineapigs and saponin fraction showed abortifacient in rabbits.

Chemical constituents

Tigogenin and diosgenin from rhizomes and stems; saponin A, m.p. 305^0, saponin B, m.p. 232^0, saponin C, m.p. 301^0 and B-sitosterol glycoside from rhizomes; rhizomes contained diosgenin (2.6%); a-amyrin stearate, B-amyrin and lupeol palmitate from leaves; effect of rhizomes diameter (5.0 cm to 15.0 cm) on sapogenin content determined; thinnest group contained largest amount of sapogenin (2.7%).

39. *Cucumis melo* Linn. (Cucurbitaceae)

Common names
Eng.: musk melon
Hindi: Kharbuza
Beng.: Kharmuj

Distribution

A creeping annual, nature to Africa and is now commonly cultivated throughout India, particularly in hot and dry climate for its fruits which are eaten.

Parts used: Fruits.

Pharmacological properties

In mice and rats ecbolic property was found.

Chemical constituents

Meloside A (6C-diglucosylapigenin), meloside L (6C-diglucosylluteolin) and their caffeoryl esters isolated from leaves, six carotenes isolated, three of which identified as a-carotene (1.82), B-carotene (94.33) and y-carotene (2.30%).

40. *Cucumis trigonus* Roxb. (Cucurbitaceae)

Common names
Hindi: Bhakura
Beng.: Gomuk
Sans.: Vishala

Distribution

An annual or perennial climber, not cultivated, distributed in West Bengal and some other regions of India.

Parts used: Fruits.

Pharmacological activities

On isolated guinea pig uterus stimulant activity has been noted on application of fruits.

Chemical constituents

The plants yield a fatty oil, Stigmast-7-en-3B-ol, its B-O-D-glucoside, alnusenone and alnusenol isolated.

41. *Cuminum cyminum* Linn. (Apiaceae)

Common names
Eng.: Cumin
Hindi: Jira
Beng.: Jira

Distribution

A herb native to the Mediterranean region and is now commonly grown in Punjab and U.P. for aromatic fruits which are used as spices and for flavoring purposes.

Parts used: Seeds.

Pharmacological activities

In female albino rats 100% antifertility effect has been shown on application of seeds.

Chemical constituents

Fruits yield an essential oil and also a fixed oil. A-pigenin-7-O-glucoside and luteolin-7-O-glucoside isolated from fruits.

42. *Curcuma domestica* Valeton. (Zingiberaceae)

Common names
Eng. Turmeric
Hindi: Haldi
Beng.: Halud

Distribution

A perennial herb cultivated mainly in Tamilnadu, Andhra Pradesh, Maharashtra, Bihar, Kerala and Orissa.

Parts used: Rhizomes.

Pharmacological activities

100% anti-fertility in female albino rats and no anti-ovulatory in rabbits also anti-fertility activity in albino rats. Petroleum ether extract showed resorption of the implants.

Chemical constituents

Analysis of Indian turmeric gave the following values; moisture 13,1, protein 6.3, fat 5.1, mineral matter 3.5, fiber 2.6, carbohydrates 69.4 and carotene calculated vitamin A, 50 I. U. 180-183^{0}C) is also obtained from turmeric.

43. *Cuscuta reflexa* Roxb. (Convolvulaceae)

Common names
Eng.: Dodder
Hindi: Akasbel
Beng.: Swarnalata

Distribution

A turning whitish yellow leafless thread-like parasitic herb growing in W. Bengal and in some other states.

Parts used: Whole plant.

Pharmacological activities

On isolated guineapig uterus stimulant activity has been noted.

Chemical constituents

Dulcitol, luteolin, quercetin and a glycoside of luteolin, m.p. 3180, isolated from stem.

44. *Cyperus esculentus* Linn. (Cyperaceae)

Common names
Eng.: Earth almond
Hindi: Chichada

Distribution

A perennial grass like sedge, indigenous to West Asia and North Africa but occurring commonly in U.P., Punjab and South India. Corms are eaten as food and chufa oil obtained from corms is used for cooking.

Parts used: Whole plant.

Pharmacological activities

In mice and rats ecbolic property has been noted.

Chemical constituents

B-sitosterol from tubers was isolated.

45. ***Cyperus rotundus*** **Linn. (Cyperaceae)**

Common names
Eng.: Nut grass
Hindi: Motha
Beng.: Muthaghas

Distribution

A perennial sedge distributed throughout India.

Parts used: Rhizomes.

Pharmacological activities

In female rats, anti-estrogenic property has been noted.

Chemical constituents

See Anti-inflammatory Chapter.

46. ***Daucus carota*** **Linn. (Apiaceae)**

Common names
Eng.: Carrot
Hindi: Gajar
Beng.: Gajar

Distribution

An annual or biennial much branched herb native to Europe and Mediterranean region and is extensively cultivated in Punjab, Haryana, U.P. and M.P. for its fleshy tap roots which are eaten raw cooked boiled.

Parts used: Seeds.

Pharmacological activities

Seeds have anti-implantation activity in mice and chloroform fraction of petroleum ether extract of seeds responded marked inhibitory effect on oxytocin induced contractions on isolated rat uterus, along with inhibited the spontaneous activity of isolated rat uterus, good anti-fertility activity in female rats, anti-progestational activity in rats also showed anti-estrogenic, anti-fertility and anti-progestational activity showed inhibitory effects on spontaneous uterine motility and oxytocin induced contractions anti-fertility activity is significant in female albino rats, aqueous extract of seeds produced inhibition of implantation in rats, petroleum ether extract of seeds inhibited implantation in albino rats, treatment of alcoholic extract of seeds in rats showed significant anti-fertility activity.

Chemical constituents

Volatile oil, an aldehyde, m.p. 112^0 and a compound m.p. 230^0, isolated from fruits; a new sesquiterpene-daucene, b.p. 960/4 mm from essential oil; analysis of oil showed presence of α-pinene, nopinene, sabinene, dipentene, p-thymol, linalool, geraninol, bergamottin, B-bisabolene a sesquiterpene alcohol, m.p. 119^0 from essential oil of wild carrot, linalool, geraniol, geranyl acetate, α- and β-pinene, sabinene, limonene, bergamotene, bisabalene, daucene, carotol daucol, p-thymol, α-carcumene asarone and elemiein from essential oil of fruits of wild carrot growing in Northern Caucasus; farnesene and B-elemene also isolated; structure of daucol; choline and a quaternary base isolated from seeds; a diglycoside of cyaniding isolated from var. sativa DC.

47. ***Deeringia amaranthoides*** **Merrill. (Amaranthaceae)**

Common names
Hindi: Latman
Beng.: Golamohani
Assam: Monbir

Distribution

Bihar, Tirhut, N. Bengal, C. Bengal.

Parts used: Whole plant excluding root.

Pharmacological activities

Spermicidal activity has been noted in human and rat semen.

48. *Dimeria gracilis* Nees. (Poaceae)

Distribution

S.E. Asia, Indoman and Australia.

Parts used: Whole plant.

Pharmacological activities

Whole plant extract showed spermicidal activity in human and rat semen, saponin from whole plant showed spermicidal activity in human semen.

49. *Dryopteris filix-Mas.* (Polypodiaceae)

Common names

Beng.: Dhenki sak.

Distribution

W. Bengal and other parts of India.

Parts used: Rhizomes.

Pharmacological activities

Rhizomes extracts in mice showed decreased fertility, the drug had some toxic effects.

Chemical constituents

Filicin, m.p.92^0, α-flavaspidic acid, m.p. 93^0, albaspidin, m.p. 145^0 and filixin acid, m.p. 168^0 from rhizomes; synthesis of filicinic acid; five acids identified in fronds-headeche-7,10,13-trienoic, octadeca-9,12,15-trienoic, eicosa-8,11,14-trienoic, eicosa-5,8,11,14-tetraenoic and eicosa-5,8,11,14,17-pentaenoic acids.

50. *Embelia ribes* Burm. F. (Myrsinaceae)

Common names

Hindi: Baberang
Beng.: Biranga
Sans.: Vidanga

Distribution

A large climbing shrub with greenish yellow flower distributed in Sikkim Himalaya foot hills ascending to 4,000 ft, Assam, Manipur-3-5000 ft., hills of Western peninsula with flowering in Nov-Feb.

Parts used: Berry fruits, seeds, roots.

Pharmacological activities

Aqueous extract of fruit in female rats showed anti-implantation activity, no toxic effects, anti-estrogenic activity in women and female albino rats, seed extract markedly inhibited estrogen induced alkaline phosphatase activity in the endometrium of immature rabbits. No anti-fertility shown in mice, also 66.6% activity on late pregnancy in female albino rats, application of embelin showed antiimplantation activity in albino rats, treatment of embelin showed 100% anti-implantation in female albino rats, anti-ovalatory activity in rabbits, no oestogenic or anti-estrogenic activity in rat, mixture with borax exhibited contraceptive effect, extracts of berry fruit showed oestrogenic activity in female albino rats, anti-fertility activity in female rats, treatment with embelin showed slight prolongation of the estrus phase of the vaginal cycle in rats and guineapigs.

Chemical constituents

A new compound vilangin (0.06%) m.p. 264^0. Identified as methylenebis-(2,5-dihydroxy-4-undecyl-3,6-benzoquinone) isolated from berries. Embelin was isolated.

51. ***Ensete superbum*** **Roxb. (Musaceae)**

Common names
Eng.: Chowani

Distribution

Distributed from Bombay to Western Ghats to Travancore hills and ravine slopes and in Assam. Tropical Africa, Madagascar, S. China, S.E. Asia, Indomal.

Parts used: Seeds.

Pharmacological activities

Showed antifertility in female albino rats. Antifertility effect of VIDR-2GD fraction of seed extract in mice and rats. Incomplete inhibitory effect of pregnancy in rabbits.

52. ***Euphorbia dracunculoides*** **Lam. (Euphorbiaceae)**

Common names
Hindi: Chagulputputi
Beng.: Jaychee
Punjab: Kangi

Distribution

Sub tropical and warm temp. regions.

Parts used: Whole plant.

Pharmacological activities

On uteri guineapig, rat and rabbits cholinergic effect was demonstrated on isolated human uterine strips, short lived spasms were encountered.

Chemical constituents

Euphorbol, m.p. 90^0, and a flavone glycoside isolated from plant. Leaves and stalk contain a glycoalkaloid, euphorbine, and the seeds, a phenolic substance besides a high percentage of fined drying oil.

53. ***Euphorbia tirucalli*** **Linn. (Euphorbiaceae)**

Common names
Eng.: Milk bush
Hindi: Konpal, Sehund, Nevli
Beng.: Lanka sij

Distribution

A native of Africa naturalized in Benga' and Western Peninsula, chiefly as a large hedge plant.

Parts used: Latex.

Pharmacological activities

On isolated non-gravid guineapig uterus, oxytocic activity was shown.

Chemical constituents

Hentriacontane, hentriacontanol, B-sitosterol, taraxerol, 3,3'-di-O-methylellagic acid and ellagic acid isolated stems.

54. ***Ferula alliacea*** **Boiss. (Apiaceae)**

Parts used: Fruits.

Pharmacological activities

Fruit extract completed inhibition of human chorionic gonadotropin. Coumarin from fruits blocked the uterotrophic responses of exogenous oestrogen. There was no effect on progesterone response of uterus. Irregularity of oestrus cycle, delay in mating, response of uterus, Irregularity of oestrus cycle, delay in mating, significant decrease in both fertile mating and in the number of offsprings, influenced the sexual maturity of immature female rats. Suppressed the action of the hormone in human chorionic gonadotrophin (HGG). Coumarin from fruit also completely annuls the biological potency of HGG.

Chemical constituents

Coumarin.

55. *Ferula asafoetida* Linn. (Apiaceae)

Common names
Eng.: Asafoetida
Hindi: Hing
Beng.: Hing
Sans. Bahlika

Distribution

A perennial herb, commonly grown in Punjab and Kashmir.

Parts used: Fruits.

Pharmacological activities

Estrogen induced alkaline phosphatase activity in the endometrium of immature rabbits could be inhibited by simultaneous administration of the plant extract, mixture of ingredients of *Embelia ribes, piper longum, Ferula assafoetida* and borax showed contraceptive effect.

Chemical constituents

The plants yield an essential oil, oleum Asafoetida, Luteolin and luteolin-7-O-B-D-glucopyranoside from fruits.

56. *Gossypium herbaceum* Linn. (Malvaceae)

Common names
Eng.: Levant cotton.

Distribution

A shrub commonly cultivated for cotton.

Parts used

Stem, stem bark, root, seeds.

Pharmacological activities

No anti-implantation activity in albino rats have been recorded. Gossypol, a phenolic compound extracted from plant, reduces sperm density to less than 4 million/ml 99.9% of men and impairs motility. Normal sperm density is restored within several months of discontinuation of the drug.

57. *Grewia asiatica* Linn. (Tiliaceae)

Common names
Beng.: Phalsa

Distributed

Commonly planted fruit tree, grown in W. Bengal and some other regions.

Parts used: Seeds.

Pharmacological activities

Significant anti-fertility activity in female albino rats have been noted, when treated with seed oil. Petroleum ether extract, alcoholic and aqueous extract of seeds responded anti-implantation activity in albino rats.

Chemical constituents

Taraxa sterol, B-sitosterol and erythrodiol isolated from bark; lupeol, lupenone, betulin and friedelin isolated from betulin, B-sitosterol, quercetin, its 3-O-glucoside, Naringenin and its 7-O-glucoside isolated from flowers; a new lactone- 3,21,24-trimethyl-5,7-dihydroxy hentriacontanoic acid δ-lactone isolated from flowers and grewinol isolated from flowers and identified as tetratriacontan-22-ol-13-one.

58. *Gypsophila cerastioides* D. Don. (Caryophyllaceae)

Common names
Eng.: Herbaceous annual.

Distribution

Egypt. Australia, New Zealand.

Parts used: Whole plant.

Pharmacological activities

Extracts of whole plant showed spermicidal activity in Human and rat semen treatment with saponin from plant showed spermicidal activity in human semen.

Chemical constituents

Saponin.

59. *Hagenia abyssinica* (Bruce) J.F. Gmelin (Rosaceae)

Common name
Eng.: Kousso

Distribution

Abyssinia to Malawi.

Parts used: Flowers.

Pharmacological activities

Flowers extracts responded decreased fertility in female mice but was highly toxic.

Chemical constituents

Four phloroglucinol derivatives – kosins k_1, m.p. 167^0, k_2, m.p. 110^0, k_3, m.p. 177^0 and k_4, m.p. 174^0 isolated as mixtures of isovaleryl and 2-methyl isobutyryl esters; structure of kosin k_1 elucidated.

60. *Hibiscus rosa-sinensis* Linn. (Malvaceae)

Common names
Eng.: Chine-rose, shoe flower
Beng.: Jaba

Distribution

Throughout India.

Parts used: Flowers.

Pharmacological activities

Alcoholic extract of flowers has significant effect on testis in albino rats, benzene extract showed antifertility activity in albino rats, alcoholic and benzene extract suppressed the oestrone induced gain in uterine weight in immature albino rats, showed anti-implantation activity in women, benzene extract distrupted the oestrus cycle in rats, reduction in ovarian uterine and pituitary weights, responded post-coital anti-fertility activity in female albino rats. Tubules spermatagenes arrested at early spermatid stage in male rats, extract of flowers also responded 80% anti-fertility activity in female rats, benzene extract of flowers failed to show estrogenic progestrogenic activity and also androgenic activities.

Chemical constituents

Quercetin-3-diglucoside, 3-7-diglucoside, cyaniding-3,5-diglucoside and cyaniding-3-sophoroside-5-glucoside isolated from deep yellow flowers; all above compounds and kaempferol-3-xylosylglucoside isolated from ivory white flowers.

61. *Hippophae salicifolia* D. Don (Elaeagnaceae)

Common names
Hindi: Chuma, Kalabis

Distribution

Europe, Asia, North America. A small deciduous tree occurring in temperate Himalayas.

Parts used: Bark.

Pharmacological activities

Degenerative changes in somniferous epithelium of young male rats have been noted. Inhibition of testosterone stimulated development of seminal vesicles in castrated rats has been observed.

Chemical constituents

A sterol glycoside, two phytosterols, three waxy compounds and a compound, m.p. 222^0, isolated from bark.

62. *Hydrocotyle javanica* Thunb. (Apiaceae)

Common names
Mar.: karing
Tam.: Vallarai
Tel.: Saraswataku

Distribution

Herb distributed in the Himalaya from Kashmir and Kashia Hills at 600 to 2000 m. also on mountains of Western Ghats, Nilgiris and Palnis.

Parts used: Whole plant.

Pharmacological activities

Produced spermicidal activity in rat semen.

63. *Hyptis suaveolens* Polt. (Lamiaceae)

Common names
Hindi: Vilayti tulsi
Beng.: Bilati tulsi
Oriya: Banga tulsi

Distribution

Throughout India.

Parts used: Leaves.

Pharmacological activities

In female albino rats 100% anti-fertility have been noted.

Chemical constituents

Isolation and structure determination of two new diterpenes-suaveolic acid and suaveolol. Plant yields essential oil containing 1-sabinene, d-limonene and azulenic sesquiterpenes. Anti-A haemagglutimin from seeds.

64. ***Impatiens duthviel*** **Hook. F. (Balsaminaceae)**

Parts used: Whole plant.

Pharmacological activities

Spermicidal activity in rat semen has been noted.

65. ***Jatropha curcas*** **Linn. (Euphorbiaceae)**

Common names
Eng.: Physic nut
Hindi: Bagbherenda
Beng.: Erandagachh

Parts used: Fruits and seeds.

Pharmacological activities

Application of dry fruits of seeds (2.3%) showed no sign of pregnancy in female rats but recorded full reproduction efficiency when fed with normal diet only.

Chemical constituents

Ash of seeds (4.38%) contained Ca, Mg, Na, K and traces of P; presence of oleic, linoleic, mynstic, plamitic, stearic, arachidic acids and sitosterol; detection of glucose, arabinose, xylose and rhamnose in seeds by PC.

66. ***Lithospermum officinale*** **Linn. (Boraginaceae)**

Common names
Eng.: Gronwell

Distribution

Herb distributed in Kashmir and Kumaon at 1500-2700 m alt.

Parts used: Whole plant.

Pharmacological activities

Reversible inhibition of estrus in mice noted when aqueous extract was employed.

Chemical constituents

Non-saponifiable matter from root of Japanese var erythrorhizon contained valeric and isovaleric acids; isolation of an octadece-4,8,12,15-tetraenoic acid from fruits; isolation of scyllitol from plant and caffeic, chlorogenic and ellagic acids and amino acids from leaves; rutin (0.54%) isolated.

67. ***Lithospermum ruderale*** **Dougl. Ex. Lehm. (Boraginaceae)**

Parts used: Roots.

Pharmacological activities

Treatment of root extract on adult female mice with regular cycles had prolonged diestrum, cold water infusion of roots induce sterility when taken daily for a period of 6 months. Follicular atresia of ovaries and mild uterine atrophy were shown when roots are given to virgin mice for one year; roots reduce ovarian stimuli in female mice also antiestrus effect in rats have been noted. The aqueous extract precipitated with a alcohol has been effective variously, influences growth retardation, adrenal hypertrophy thymus involution and atrophy of the sex organs along with cessation of estrus, effective in normal and adrenalectonized rats, indicating that the action is not mediated through adrenals.

68. *Madhuca butyracea* Roxb. (Sapotaceae)

Common names
Hindi: Phulwara
Kumanun: Bhalel

Distribution

Distributed in U.P. and Bihar.

Parts used: Seeds.

Pharmacological activities

Spermicidal activity in human and rat semen has been noted, saponin from seeds induce spermicidal activity in human semens.

Chemical constituents

A sterol glucoside, m.p. 276^0 and a flavonoid, m.p. 222^0 from nuts; proanthocyanidins consisting of leucocyanidin and oligosaccharide units (R-O-xylose arabinose-rhamnose-glucose) isolated; α-spinasterol and B-sitosterol-B-D-glucoside, a- and B-amyrin acetates from bark and fruit pulp.

69. *Maesa indica* Roxb. (Myrsinaceae)

Common names
Beng.: Ramjani
Assam: Awuapat
Mar.: Atki

Distribution

India and Nepal.

Parts used

Whole plant excluding root.

Pharmacological activities

Spermicidal activity in rat semen has been noted.

Chemical constituents

Sitosterol and quercetin-3-rhamnoside isolated from leaves.

70. *Mallotus philippensis* Lam. (Euphorbiaceae)

Common names
Eng.: Kamala tree
Hindi: Kamala
Beng.: Kamala

Distribution

India and Nepal.

Parts used

Seeds, glands and hairs of the capsule of fruit.

Pharmacological activities

Whole plant showed anti-fertility activity in rats, seeds have no anti-fertility activity in albino rats and mice, glands and hairs from the capsular fruits have some effect to produce infertility in mice, the oestrus cycle was so disturbed that animals fail to mate, anti-fertility effect of the active substance was due to counter-action of the effect of chorionic gonadotropin also reduced the fertility of male and female rats and guineapigs.

Chemical constituents

Betulin-3-acetate, lupeol, lupeol acetate, sitosterol and bergenin isolated from heartwood; acetylaleuritolic acid, α-amyrin, sitosterol and bergenin isolated from bark. Rottlerin and acetyl rottbrin from the plant extract.

71. ***Medicago sativa*** **Linn. (Fabaceae)**

Common names
Eng.: Alfalfa
Hindi: Lasunghas
Punjab: Lusan

Distribution

A perennial herb of temperate Europe, Asia and North Africa widely cultivated as fodder for live stock.

Parts used: Whole plant.

Pharmacological activities

Coumestrol from plant increases the age of maturity and depresses the egg production in white leghorn pullent.

Chemical constituents

A highly toxic saponin isolated and characterized as triglucoside of medicagenic acid; detection of monogalactosyl (3.2), digalactosyl (7.7) and sulfoquinovosyl (0.8%) diglycerides in glyceroglycolipids from leaves by GC; a saponin isolated from roots yielded on hydrolysis hederagenin, glucose and arabinose; myrcene, limonene and linalool isolated from flowers; a new isofavon-sativin, m.p. 125^0 isolated from leaves; medicarpin-B-D-glucoside isolated from roots; coumestrol isolated; benzoyl mesotartaric acid and benzoyl (s)-(-) malic acid isolated; a-tocopherol found predominant tocopherol isomer in alfalfa concentrate.

72. ***Mentha arvensis*** **Linn. (Lamiaceae)**

Common names
Eng.: Field mint
Hindi: Podina
Beng.: Pudina

Distribution

West Bengal and other parts of India.

Parts used: Leaves.

Pharmacological activities

Significant antifertility activity in female albino rats has been noted and no post-coital antifertility activity in other experiments. Alcoholic extract showed 80-100% anti-implantation activity in female rats, experiment on rabbits showed anti-ovulatory activity. Aqueous extract responded 60% anti-implantation activity where as petroleum ether extract has abortifacient activity and alcoholic extract responded 80% anti-implantation activity in female albino rats.

Chemical constituents

The plants yield an essential oil. Acacetin, apigenin, diosmetin, eriodictyol, hesperetin and luteolin isolated from aerial parts.

73. ***Michelia champaca*** **L. (Magnoliaceae)**

Common names
Hindi: Champa
Beng.: Champa
Sans.: Champaka

Distribution

Wild in the Eastern Himalayas and North-East India up to 900 m. Western Ghats and South India, widely cultivated in various regions of India.

Parts used: Bark.

Pharmacological activities

Bark is employed as an abortifacient for 2-3 months' old pregnancy. A roof paste (roots

6-8 mm long) with black pepper is employed as an oral contraceptive given for 3 days after menstruation.

Chemical constituents

See Anti-inflammatory Chapter.

74. *Momordica charantia* Linn. (Cucurbitaceae)

Common names
Eng.: Bitter gourd
Hindi: Karela

Distribution

A climbing herb found throughout India, often cultivated for fruits which are used as vegetables.

Parts used: Roots, leaves.

Pharmacological activities

Roots showing uterine stimulant activity on isolated guineapig uterus. Aqueous extract of leaves shows no significant anti-fertility and estrogenic activity in pregnant rats.

Chemical constituents

See Antidiabetic Chapter.

75. *Moringa oleifera* Lam. (Moringaceae)

Common names
Beng.: Sajina
Hindi: Shajnah
Sans.: Sobhajana

Distribution

Throughout India.

Parts used: Root, Stem-bark.

Pharmacological activities

Roots have effect in anti-fertility activity on late pregnancy in rats, stem-bark has little oxytoxic activity on isolated guineapig uterus, but no anti-fertility activity in albino mice and rats .

Chemical constituents

One antibiotic named ptergospermin was isolated from plant root. It is reddish brown oil and is most active at pH of 5 and activity decreases as the pH approaches 8.

76. *Morus alba* Linn. (Moraceae)

Common name
Eng.: White mulberry
Hindi: Tut
Beng.: Toot

Distribution

Cultivated throughout the plain parts of India.

Parts used: Leaves.

Pharmacological activities

Leaves have ecbolic property in mice and rats.

Chemical constituents

Mulberrin, m.p. 153^0, mulberrochromene, m.p. 232^0, cyclomulberrin, m.p. 233^0, characterized from stems, roots and bark.

77. *Ocimum americanum* Linn. (Lamiaceae)

Syn. **O. canum Sims**

Common names
Eng.: Hoary Basil
Hindi: Kalitulsi
Sans.: Ajaka

Distribution

Throughout the tropical regions of the world.

Parts used: Seeds.

Pharmacological activities

In women, 90% antifertility activity (along with Actiniopteris radiata) has been noted.

Chemical constituents

Polysaccharide contained xylose, arabinose, rhamnose and galactose, galacturonic acid and glucuronic acid detected by PC.

78. *Ocimum sanctum* Linn. (Lamiaceae)

Common names
Eng.: Sacred Basil
Hindi: Tulsi
Beng.: Tulsi

Distribution

A strongly scented shrub. It is cultivated throughout India, and also considered as a sacred plant to the Hindus. These plants are of two types, one is purple type, called Krishna tulsi and another is green called Sri tulsi.

Parts used: Leaves.

Pharmacological activities

Benzene extract of leaves responded 80% anti-fertility activity in female rats but petroleum ether extract responded 60% anti-fertility activity in female rats also anti implantation activity in albino rats has been noted.

Chemical constituents

See Hepatoprotective Chapter.

79. *Ougeinia dalbergioides* Benth. (Fabaceae)

Common names
Eng.: Chariot tree
Hindi: Sandan
Beng.: Tinis

Distribution

Throughout India and Nepal.

Parts used: Stem bark.

Pharmacological activities

Stem bark extract responded spermicidal activity in rat semen.

Chemical constituents

Bark contains tannin (7%). The heartwood contains homoferreirin (0.4%) and a new isoflavone, ougenin.

80. *Oxalis corniculata* Linn. (Oxalidaceae)

Common names
Eng.: Indian sorrel
Hindi: Amrulsak
Beng.: Amrul sak

Distribution

Throughout India and Nepal.

Parts used: Whole plant.

Pharmacological activities

Crude drug extract showed estrogenic activity in female albino rats.

81. *Paeonia emodi* Wall. (Paeoniaceae)

Common names
Eng.: Himalayan Peony
Hindi: Ud-salap
Punjab: Mamekh

Distribution

Bihar, Punjab and Kashmir.

Parts used: Tubers.

Pharmacological activities

Tubers have marked uterotonic activity in isolated uteri of rat, guineapig, rabbit and dog *in situ*.

Chemical constituents

Starch (9.5), sucrose (5.4), malic acid (0.47), oxalic acid (0.36), tartaric acid (0.34%) and benzoic acid present in roots.

82. *Petroselinum hortense* Hoffm. (Apiaceae)

Common names
Eng.: Parsley
Kan: Achu mooda

Distribution

A herb, native of Europe, now cultivated throughout India.

Parts used: Whole plant.

Pharmacological activities

Ecbolic property in mice and rats have been noted.

Chemical constituents

Apiole (1-allyl-2,5-dimethoxy-3,4-methylene-dioxybenzene), allyltetramethoxybenzene, myristicin, a mixture of flavone glycosides and a compound, m.p. 145^0 from seeds; 1-allyl-2,3,4,5-tetramethoxybenzene, apigenin-7-apioglucoside and luteolin-7-apioglucoside from fruits; a new syntheses of apigenin, m.p. 346^0.

83. *Pimpinella anisum* Linn. (Apiaceae)

Common names
Eng.: Aniseed
Hindi: Saunf
Beng.: Muhuri, mithajira

Distribution

Africa, North America and South America.

Parts used: Seeds and roots.

Pharmacological activities

Oil extracted from seeds and roots has oestrogenic activity in rats.

Chemical constituents

An aliphatic alcohol, m.p. 74^0, an essential oil, a fixed oil and mannitol from seeds; myricanol and mannitol from roots.

84. *Pimpinella diversifolia* DC. (Apiaceae)

Distribution

Decan.

Parts used: Whole plant.

Pharmacological activities

Plant extract responded spermicidal activity in rat semen.

Chemical constituents

Ammirin (isoangenomalin) and oxypeucedanin isolated from aerial parts.

85. *Piper aurantiacum* Wall. (Piperaceae)

Common names
Hindi: Shambhaluka bij
Beng.: Renuk

Distribution

W. Bengal.

Parts used: Fruits.

Pharmacological activities

Oxytocic activity has been observed by alkaloidal fraction. The defatted tannic free alcoholic extract has shown parasympathomimetic activity.

Chemical constituents

Piperine, piperettine, sylvatine and B-sitosterol isolated from fruits. A new amide-aurantiamide and its acetate isolated from seeds. Stearic and linoleic acids, triacontane, cholesterol, cholestanol also isolated from fruits.

86. *Piper betle* Linn. (Piperaceae)

Common names
Eng.: Betel
Beng.: Pan
Hindi: Tambula

Distribution

Bengal, Orissa, U.P, Bihar and Assam.

Parts used: Roots.

Pharmacological activities

Roots extracts have anti-fertility activity.

Chemical constituents

Beta Sitosterol from the root and also from the leaf. Leaves contain starch, sugars; tannin, diastases (0.8 to 1.8%) and essential oil (Betel oil) to the extent of even 4.2% in some leaves. The essential oil is a light yellow liqid of aromatic odour and sharp burning teste. The specific gravity varies from 0.958 to 1.057.

87. *Piper chaba* Hunter. (Piperaceae)

Common names
Eng.: Java long pepper
Hindi: Chab
Beng.: Chai

Distribution

North-East Himalaya.

Parts Used: Fruits.

Pharmacological activities

Fruits have anti-fertility activity.

Chemical constituents

A new amide-2,4, 12-trienamide isolated from fruits.Filfiline (N-isobutyldocosa-transe-2, transe-4 CIS-10 trienamide) also isolated from fruits.

88. *Piper longum* Linn. (Piperaceae)

Common names
Eng.: Long pepper
Hindi: Pipal
Beng.: Piplamur

Distribution

Cultivated in W. Bengal, Karnataka and Tamil Nadu.

Parts used: Roots, leaves and fruits.

Pharmacological activities

Mixtures of roots and leaves along with ingredients of *Embelia ribes* and *Ferula assafoetida* responded contraceptive responses; fruits of *P. longum* have inhibitory effect. Extracts showed anti-fertility activity in rats.

Chemical constituents

Piperlongumine, piperlonguminine piperine, sisamine, methyl-3,4,5-trimethoxycinnamate isolated from roots.

89. *Pisum sativum* Linn. (Fabaceae)

Common name
Eng : Pea
Hindi: Matar
Beng.: Matar

Distribution

W. Bengal, Bihar, U.P and some other states.

Pharmacological activities

Seeds have contraceptive action in albino rats and only meta-xylo-hydroquinone (not its isomers) interfered with progesterone level in blood single dose (0.1 mg) of the active principle responded abortion, reabortion but the substance did not prevent nidation (140), ecbolic property in mice and rats. It has been observed that application of meta-xylo-hydroquinone to mature female rats control condition, before mating and during frits trimester of gestation, failed to delay mating, also prevent implantation. Seed oil showed temporary sterility produced in both sexes of rats.

Chemical constituents

D-Galacturonyl-L-rhamnose isolated from hydrolysate of an acidic polysaccharide isolated from seeds. A celebroside isolated from seeds which on hydrolysis yielded hydroxytricosanoic acid, sphingosine lyase and glucose.

90. *Plumbago indica* Linn. (Plumbaginaceae)

Common names
Hindi: Chitra
Beng.: Lal-Chitra
Sans.: Chitrak

Distribution

Throughout India.

Parts used: Roots.

Pharmacological activities

Responded weak anti-fertility activity in late pregnancy in rats also showed oxytoxic action on isolated uterus of rat, guineapig and human.

Chemical constituents

The root bark contains an orange-yellow pigment plumbagin (2-methyl-5-hydroxy-1,4-naphthoquinone, $C_{11}H_3O_3$ m.p. 77-78^0) a sitosterol glycoside ($C_{33}H_{56}O_6$), m.p.

259-260^0 a sitosterol, a fatty alcohol, probably arachidyl alcohol, tannin and amorphous brown pigment. The flowers contain 3-rhamnoside of pelargonidin, cyanidin, delphinidin and kaempferol.

91. ***Plumbago zeylanica*** **Linn. (Plumbaginaceae)**

Common names
Hindi: Chita
Beng.: Chita
Sans.: Chitraka

Distribution

Gangetic plains of India.

Parts used: Root and fruits.

Pharmacological activities

Alcoholic extract (50%) responded 100% anti-implantation in rats but no anti-implantation activity on rabbits.

Chemical constituents

Plumbagin, 3-chloroplumbagin and a new substance 3,3'-biplumbagin, a new binaphthaguinone-chitranone together with zeylinone, isozeylinone, equiptinone and droserone isolated from roots.

92. ***Plumeria acutifolia*** **Poir. (Apocynaceae)**

Common names
Eng.: Temple tree
Hindi: Golainchi
Beng.: Dalan phul

Distribution

Throughout India.

Parts used: Bark.

Pharmacological activities

Responded uterine stimulation in isolated guineapig uterus.

Chemical constituents

Fulvoplumierin, m.p. 147^0, plumericin, m.p. 211^0 and three new compounds iso plumericin, m.p. 200^0, Beta dihydro-plumericin, m.p. 191^0 and Beta dihydro-plumericinic acids, m.p. 189^0 from roots; fulvopulmierin, Beta lupeol and plumieride from stem bark.

93. ***Polygonum hydropiper*** **Linn. (Polygonaceae)**

Common names
Eng.: Pepper-Wort
Beng.: Packur-mul, panimaricha

Distribution

Throughout India, ascending to an altitude of 7.000 ft in the Himalayas.

Parts used: Roots, leaves, whole plant.

Pharmacological activities

Roots responded sterility in female guineapig, no evidence of estrogenic of androgenic activity. In other experiments 66.6% anti-fertility activity on early pregnancy in rats, petroleum ether extract responded anti-ovulatory activity in rabbits and anti-implantation in albino rats but the whole plant impaired the fertility of male and female mice and produced temporary sterility in guineapig.

Chemical constituents

The herb contains formic acid, acetic acid and baldrianic acid, much tannin and small amount of essential oil, the root contains oxymethyl-anthraquinones.

94. *Psoralea corylifolia* Linn. (Fabaceae)

Common names
Eng.: Bakuchi
Hindi: Babchi
Beng.: Bavachi
Sans.: Bakuchi

Distribution

Throughout India.

Parts used: Seeds.

Pharmacological activities

Fertility of adult female rats was impaired by seed extract.

Chemical constituents

See Anti-inflammatory Chapter.

95. *Punica granatum* Linn. (Punicaceae)

Common names
Eng.: Pomeranate
Hindi: Anar
Beng: Dalim

Distribution

Small tree generally cultivated throughout India.

Parts used: Fruit skin.

Pharmacological activities

Fertility of female rats and guineapigs was reduced by fruit skin extract.

Chemical constituents

The plants contain iso-pelletierine. Sitosterol and ursolic acid isolated; two tannins-punicalagin and punicalin isolated from peels and their structures determined, hydrolysis of punicalagin yielded ellagic acid and punicalin, the latter yielded glucose and a tetralactone; pectin isolated from fruits contained mannose, galactose, rhamnose, arabinose and glucose in ratio of 1:1:3: 2.1:4:4:6:3; principle sugar acid was galacturonic acid.

96. *Randia dumetorum* Poir. (Rubiaceae)

Common names
Eng.: Common emetic
Hindi: Mainphal
Beng.: Mainphal

Distribution

India and Nepal.

Parts used: Fruits, seeds and pulp.

Pharmacological activities

Ethanolic extract of fruits and seeds shows no anti-implantation activity in rats but anti-ovulatory effect in rabbits has been observed pulp showed uterine stimulant activity on isolated guineapig uterus. Anti-implantation activity in albino rats has been observed with the treatment of oleanolic acid 3-B-glucoside isolated from seeds.

Chemical constituents

Oleanolic acid 3-Beta glucoside.

97. *Rauwolfia serpentine* Benth. (Apocynaceae)

Common names

Eng.: Serpentine
Hindi: Chandrabhaga
Beng.: Chandra, Sarpa-gandha

Distribution

A small shrub found from Himalayas southwards to Peninsular India.

Parts used: Roots.

Pharmacological activities

Roots responded decreased reproduction capacity in rats and release of pituitary gonadotrophin (hence ovulation) backed in rats and mice has been observed.

Chemical constituents

Roots contain reserpine. Structure elucidation of reserpine and deserpidine essential oil (0.22%) from roots yielded chief terpene constituent-serpoterpine, m.p. 131^0; isolation of reserpine, m.p. 360^0, Ajmalicine, m.p. 158^0, serpentine, m.p. 155^0, serpentinine, m.p. 265^0 and ajmalicine (δ-yohimbine), m.p. 250^0, from roots; reserpine (mixture of serpentine, serpentinine, ajmaline and 2 unknown compounds, free of reserpine), resajmaline (mixture of serposterol and unsaturated higher alcohols), rescinnamine and Ajmalicine (mixture of reserpine, reserpinine), yohimbine, ajmaline, serpentine by PC; serpentinine characterized as 3-hydroxyserpentine; absolute stereochemistry of ajmaline and isoajmaline; raugalline isolated and identified as ajmaline; detection of raunatine in root extract by PC; stereochemical studies in structure elucidation of yohimbine and reserpine.

98. *Rubus ellipticus* Sm. (Rosaceae)

Common names

Hindi: Hinsalu, anchhu

Distribution

From Punjab to Assam extending southwards in the Western Ghats.

Parts used: Plant excluding roots.

Pharmacological activities

Alcoholic extract responded anti-implantation activity in rats but no anti-ovulatory activity in rabbits has been observed.

99. *Salix babylonica* Linn. (Salicaceae)

Common names

Eng.: Weeping willow
Tel.: Attuppalia
Punjab: Bisa

Distribution

Throughout the temperate regions of the world. Native of Central China, now cultivated in northern parts of India.

Parts used: Plant excluding root.

Pharmacological activities

Anti-implantation in rats has been shown by alcoholic extracts but no antiovulatory activity has been noted in rabbits.

Chemical constituents

Bark and leaves are reported to contain 3-9% and 4.9% tannin. Leaves contain delphinidin and cyanidin, fragilin, salicin salicortin, salidroside, tremuloidin, triandrin and vimalin.

100. *Salix tetrasperma* Roxb. (Salicaceae)

Common names
Eng.: Indian willow
Hindi: Bod, Jalmala
Oriya: Baisi
Beng.: Panijama, Boishakhi

Distribution

Throughout the temperate regions of the world.

Parts used: Stem, bark.

Pharmacological activities

Alcoholic extract of stem bark responded anti-implantation activity in rats but failed to respond antiovulatory activity in rabbits.

Chemical constituents

Bark is reported to contain 6.5% tannin. Analysis of sun-dried nature leaves gave ash 10.05; calcium 2.71; carbon 45.006; and nitrogen 2.07%.

101. *Samanea saman* Merrill. (Mimosaceae)

Distribution

Chotanagpur.

Parts used: Plant excluding root.

Pharmacological activities

Saponin from whole plant (excluding root) responded spermicidal activity in human semen also whole plant (excluding root) has spermicidal activity in human and rat semen.

Chemical constituents

Octacosanoic acid, lupeol, α-spinasterol, α-spinasterone and lupenone isolated from bark; a new saponin-samanin β-shown to be constituted of acacic acid and glucose arabinose, xylose and rhamnose present in molar ratio of 4:2:1:1; another new saponin-samanin c,m.p. 146^0 isolated from wood and found to contain acacic acid along with glucose, arabinose, xylose, fucose and rhamnose in molar ration of 1:2:3:4; a new saponin-samanin D isolated from flowers contained acacic acid along with glucose, arabinose, xylose and rhamnose in ratio of 5:4:3:1.

102. *Sapindus mukorossi* Gaertn. (Sapindaceae)

Common names
Hindi: Ritha
Eng.: Ritha
Sans.: Urista, phenila

Distribution

Decan, Soapnut tree of N. India.

Parts used: Fruit.

Pharmacological activities

Saponin from fruit responded spermicidal activity in human semen fruit responded spermicidal activity in human and rat semen.

Chemical constituents

Two saponins-sapindoside A, m.p. 214^0 and sapindoside B m.p. 276^0 isolated and characterized as hederagenin-3(a-L-arabino-pyranposyl-2-a-L-Rhamnopyranoside) and hederagenin-3-(a-Larabinopyranosyl-2-a-L-rhamno-pyrano-syl-3-Beta D-xylopyrano-side) respectively; isolation and characterization of sapindoside C, m.p. 235^0, isolation and structure of sapindoside D; isolation and structure of sapindoside E; quercetin and kaempferol identified; stigmasterol, its glucopyranoside, 28-norolean-12-en-3-B,17-Beta diol and 2,23-ethylidenehederagenin isolated from pericarp of fruits; lauric, palmitic, oleic, stearic and arachic (arachidic) acids also detected.

103. ***Sapindus trifoliatus*** **Linn. (Sapindaceae)**

Common names
Eng.: Soapnut-tree
Hindi: Reetha
Beng.: Bararitha

Distribution

Decan.

Parts used: Pulp and seeds.

Pharmacological activities

Pulp and seeds have uterine stimulant activity on isolated guineapig uterus seeds have anti-implantation activity in female rats.

Chemical constituents

Dried fruit contains 11.5% saponin. A new saponin emerginatoside ($C_{47}H_{76}O_{11}$) m.p. 210-120 has been isolated from the aqueous extract of fruits. Kernels contain 45.4% oil. The oil has the following fatty acid composition; palmitic 5.4; stearic, 8.5; arachidic, 20.7; behenic, 2.1; oleic, 55.1; and linoleic 8.2%.

104. ***Schefflera capitata*** **Harms. (Araliaceae)**

Distribution

Nilgiri hills, tropical and sub-tropical regions of the world.

Parts used

Whole plant excluding root.

Pharmacological activities

Saponin from plant showed spermicidal activity treatment with whole plant (excluding root) also responded spermicidal activity in human and rats, saponin from plant (excluding root) showed spermicidal activity in human semen.

Chemical constituents

A new saponin scheffleraside isolated which on hydrolysis yielded echinocystic acid, fucose, galactose and glucuronic acid equimolar ratio.

105. ***Semecarpus anacardium*** **L.F. (Anacardiaceae)**

Common names
Hindi & Beng.: Bhela
Tel.: Bhallataki

Distribution

Throughout hotter parts of India, also in the foot hills of the Himalayas up to 1100 m.

Parts used: Fruit (nut).

Pharmacological activities

Nut is used in abortion of 3-5 months pregnancy.

Chemical constituents

See Anti-inflammatory Chapter.

106. ***Sesamum indicum*** **Linn. (Pedaliaceae)**

Common names
Eng.: Sesame
Hindi: Til
Sans.: Tila

Distribution

Cultivated throughout India.

Parts used: Seeds.

Pharmacological activities

Estrogenic effect in female albino rats has been noted.

Chemical constituents

A flavonoid glucoside-pedalin (0.3%), m.p. 254^0 isolated from leaves.

107. ***Sesbania sesban*** **L. (Fabaceae)**

Common names
Eng.: Egyptian Rattle Pod
Hindi: Jainti
Beng.: Jainti

Distribution

Throughout India.

Parts used: Flowers, leaves.

Pharmacological activities

Leaves have no antifertility activity in albino rats and mice. Flowers have variety of effects, it causes antifertility effect in mice and rats and abortifacient in mice also anti-fertility response in albino rats and mice has been observed.

Chemical constituents

Determination of fatty acid composition of seed oil. Leaves contain good amount of protein, calcium and phosphorus. At least six flavonols, magnesium and traces of iron are present. Vitamin C content is reported to be 89.4 mg/100 mg. Seed extract (with petroleum ether yields an oil, Oil contains palmitic, 9.0′ stearic, 17.5; lignoceric, 1.9; oleic, 24.4; linoleic, 36.3; linolenic 10.9; acids.

108. ***Solanum khasianum*** **C.B. Clarke. (Solanaceae)**

Common names
Beng.: Bon-Begun

Distribution

North-east Himalaya.

Parts used: Whole plant.

Pharmacological activities

Antiovulatory effect in rabbits has been observed when treated with alcoholic extract.

Chemical constituents

Solasonine, m.p. 276^0. Solasodine and solakhasianin, m.p. 251^0 isolated from berries.

109. ***Solanum xanthocarpum*** **Schrad. & Wendl. (Solanaceae)**

Common names
Eng.: Yellow Berried Nightshade
Hindi: Kateli Kantakari
Sans: Schrad

Distribution

Throughout W. Bengal.

Parts used: Whole plant.

Pharmacological activities

Whole plant responded spermicidal activity in rat semen stem; leaves have no anti-fertility effect in rats and guineapig. Fruits also have no anti-fertility activity.

Chemical constituents

Contains 1% of alkaloids, can from a source for cortisone and sex hormone preparation. A new sterol-carpesterol characterized as (22R) 22-hydroxy-6-oxo-4-α-methyl-5-α-stigmast-7-en-3-B-yl benzonte; scopoletin, esculin and esculetin isolated; solasodine, solasonine, solamargine and B-solamargine in fruits of Nepalese plant; two new sterols-norcarpesterol (22-hydroxy-6-oxo-4-a-methyl-24-methylcholest-7-en-3-B,22-diol isolated; dry fruits contained acids and caffeic acids (0.035%); quercetin-3-O-B-D-gluco-pyranosyl-mannopyranoside isolated together with apigenin and sitosterol.

110. ***Solidago virgaurea*** **Linn. (Asteraceae)**

Common names
Eng.: Woundwort

Distribution

Plants and Himalayan tracts of Bengal.

Parts used: Whole plant, saponin from whole plant.

Pharmacological activities

Whole plant responded spermicidal activity in human and rat semen.

Chemical constituents

Quercitrin, rutin, isoquercitrin, astragalin and kaempferol rhamnoglucoside isolated.

111. ***Spondias cytherea*** **Sonn. (Anacardiaceae)**

Common names
Eng.: Great Hog-plum, Golden apple
Beng.: Bilati Amra

Distribution

Common in West Bengal.

Parts used: Bark.

Pharmacological activities

Bark showed anti-fertility effect on mice and pigs.

Chemical constituents

Analysis of fruits gave moisture, 59.65' protein, 0.80; fat 1.79; sucrose, 8.05; crude fibre, 3.60; and ash 0.65%. Content of carotene in the fruit is 0.26 mg/100 g. Fruits are also good sources of iron. Tree exudes a gum composed of dgalactose, l-arabinose, d-xylose, a mono-O-methyl glucuronic acid and traces of l-rhamnose and I-focose.

112. *Stephania hernandifolia* Willd. (Menispermaceae)

Common names
Eng.: Tape-vine
Beng.: Akanadi
Sans.: Vanatiktika

Distribution

W. Bengal, Orissa, Assam.

Parts used: Rhizomes.

Pharmacological activities

Rhizomes have fertility promoting as well as anti-fertility properties in mice and rats. In other experiments anti-fertility effect in albino mice and rats have been observed.

Chemical constituents

Alkaloid aknadine. Steponine isolated; protostephanine, m.p. 72^0, obtained chemically from bromo protostephanine; fangchinoline, m.p. 238^0, dl-tetrandrine, m.p. 257^0, a-tetrandrine, m.p. 217^0 and d-isochondrodendrine, m.p. 309^0 isolated from roots; an amorphous alkaloid as O-acetyl derivative, m.p. 113^0, from aerial parts; a new alkaloid hernandoline (aknadinine) from aerial parts; a new alkaloid hernandoline (aknadinine) from herb; new alkaloids aknadine, m.p. 133^0 aknadinine, m.p. 70^0 aknadicine, m.p. 156^0 and a compound m.p. 66^0 from roots and rhizomes; estimation of aknadine in roots and rhizomes by TLC; isolation and structure elucidation of new alkaloid-4-demethyl hasubanonine (aknadinine) from roots; crystal structure of aknadinine determined.

113. *Taxus baccata* Linn. (Taxaceae)

Common names
Eng.: Common yew
Hindi: Thuno
Beng.: Birmi

Distribution
W. Bengal, Bihar, U.P. and Kashmir.

Parts used: Leaves.

Pharmacological activities

Leaves have various effects on reproductive cycle. It responds anti-implantation activity in albino rats, aqueous extract shows anti-implantation activity in albino rats also anti-ovulatory activity in rabbits, no uterine stimulant activity in isolated guineapig uterus, the same effect in rats and guineapigs. When the drug was mixed with red ochre, female albino rats showed some antifertility effects.

Chemical constituents

Chemical studies on taxine, isolated from leaves; a flavonoids m.p. 294^0, isolated from leaves, similar in many respects to sciadopitysin; two new biflavones isomers of sciadopitysin and sotetsuflavone m.p. 212^0 and 300^0, respectively isolated from leaves; both compounds yielded same dmethyl derivative; rhodoxanthin and eschscholtzxanthone from fruits; Beta sitosterol a methoxy triterpene, m.p. 219^0 and a compound D, m.p. 161^0 from root bark; Beta Sitosterol also isolated from wood, bark and leaves.

114. *Terminalia arjuna* Roxb. (Combretaceae)

Common names
Eng.: Arjun

Beng.: Arjhan
Hindi: Maruthu, Kahu, Arjuna

Distribution

Throughout West Bengal and some other regions in India.

Parts used: Bark.

Pharmacological activities

In female rats antifertility effect has been observed.

Chemical constituents

Arjunolic acid, tomentosic acid, Beta Sitosterol, ellagic acid, (+) leucodelphinidin and a saponin, m.p. 87^0, saponin on hydrolysis yielded arjunolic acid and glucose; structure of arjunolic acid confirmed by synthesis.

115. ***Terminalia bellirica*** **Roxb. (Combretaceae)**

Common names
Eng.: Belleric myrobalan
Hindi: Bahera
Beng.: Bhairah

Distribution

Throughout India.

Parts used: Fruits.

Pharmacological activities

Fruits have spermicidal activity in rats' semen saponin from fruit showed spermicidal activity in human semen.

Chemical constituents

A new cardiac glycoside bellericanin isolated which yielded glucose and galactose. Heartwood, bark and fruits contain ellagic acid and the seed-coat of the fruit contains gallic acid.

116. ***Trifolium alexandrinum*** **Linn. (Fabaceae)**

Common names
Hindi: Berseem
Eng.: Egyptian clover, Barseem

Distribution

An important forage crop in Punjab, Kangra in H.P. and Western U.P, also cultivated in small scale in M.P., Bihar, Maharashtra, Tamil Nadu and Karnataka.

Parts used: Seeds.

Pharmacological activities

In rats oestrogenic activities have been observed.

Chemical constituents

1-Ascorbic acid and reducing sugar content during growth and successive cuttings. Biochanin A (0.013), genistein (0.025%) and formononetin found in plant; coumestrol also isolated.

117. ***Trigonella foenum-graecum*** **Linn. (Fabaceae)**

Common names
English: Fenugreek
Hindi: Methi
Beng.: Methi

Distribution

Throughout India.

Parts used: Seeds, saponin from seeds.

Pharmacological activities

Causes ecbolic effect in mice and rats in other experiments aqueous extract causes immobilization of human and bovine speronatozoa and spermicidal activity in human and rat semen has been observed, saponin from seed showed spermicidal activity in human semen.

Chemical constituents

Two flavonoids glycosides, quercetin and luteolin, and two steroidal saponins from seeds identified by PC.

118. *Uraria lagopodioides* Desv. (Fabaceae)

Common names
Hindi: Pithavan
Beng.: Golak Chakulia
Tel: Kolaponna

Distribution

W. Bengal, Bihar, Maharashtra and Palni Hills.

Parts used: Whole plant.

Pharmacological activities

Aqueous extract of the plant showed anti-implantation activity in albino rats.

119. *Vitex negundo* Linn. (Verbenaceae)

Common names
Hindi: Sambhalu
Beng: Nishinda
Sans.: Nirgundi

Distribution

Throughout India.

Parts used: Roots and seeds.

Pharmacological activities

Roots have less anti-fertility activity in rats but seeds have significant anti-ovulatory activity in rats.

Chemical constituents

n-Tritriacontane, n-hentriacontane, n-nonacosane, Beta sitosterol, p-hydroxybenzoic acid and 5-oxyisophthalic acid from seeds; 3,4-dihydroxy benzoic acid and 5-oxyisophthalic acid from seeds; 3,4-dihydroxy benzoic acid also isolated; vanilic and p-hydroxybenzoic acids and luteolin isolated from bark; two new leucoanthocyanidins isolated from stem bark and their structures determined as 6,8-di-O-methyl-leucodelphinidin and 3′, 4′ –dl-O-methyl-leucocyanidin-7-O-rhamnoglucoside.

120. *Withania somnifera* Dunal. (Solanaceae)

Common names
Eng.: Winter cherry, Ashwagandha
Hindi: Asgand
Beng: Shavaganda

Distribution

Under shrub, found in drier parts India.

Parts used: Roots and tuber roots.

Pharmacological activities

Roots have infertility activity in mice and did

not completely abolish oestrus or mating but it delayed the processes. Roots have also effect to produced infertile mating and caused a decrease in litter size tuber roots have no uterine stimulant activity on isolated guineapig uterus.

Chemical constituents

The water soluble extract of root contains black resin which contained Hentriacontane, $C_{13}H_{64}$,a phytosterol, $C_{27}H_{46}$ O(m.p.135-36^{0}C, a mixture of fatty acids consisting of palmitic, stearic, cerotic, oleic acids etc. A C-28 steroid isolated from roots and identified as, 5,20a-dihydro-6a-epoxy-1-oxowitha-2,24-dienolide (withanolide); nine new steroidal lactones-withanolides E,F,G, H,I,J,K,L and M-isolated from leaves; seven of these characterized as 20-hydroxy-1-oxo-20R, 22R-with α-2,5,8 (14), 24-tetraenolide (withanolide H), 20-hydroxy-1-oxox-20R, 22R-with a, 3,3,8 (14),24-tetraenolide (withanolide L), 17, 20-dihydroxy-1-oxo-20S,20R-with a-3,5,8(14),24-tetraenolide (withanolide K), 17,20-dihydroxy-1-oxo-14,15a-epoxy-20S, 22R-with a-2,5,24-tetraenolide (withanolide M); another withanolide-WS-isolated from seeds.

121. *Woodfordia fruticosa* Kurz. (Lythraceae)

Common names

Eng.: Fire-Flame bush
Hindi: Dhawi
Beng.: Dhai

Distribution

Throughout India.

Parts used: Flowers.

Pharmacological activities

Abortifacient effect in mice has been observed.

Chemical constituents

Ellagic acid, polystachoside, myricetin-3-galactoside and pelargonidin-3,5-diglucoside isolated from leaves and flowers. Cyanidin-3,5-diglucoside isolated from flowers; octacosanol, B-sitosterol and chrysophanol-8-O+B-D-glucopyranoside isolated from flowers.

❑❑❑

5

MEDICINAL PLANTS FOR HEPATOPROTECTIVE

INTRODUCTION

Hepatoprotective

Liver is the principal organ for maintaining the body's internal environment. The liver has a major influence on the flow of nutrients to the rest of the body as it controls the metabolism of carbohydrates, proteins and fat. Liver is also the biggest reticulo-endothelial organ in the body and as such has important immune functions in maintaining body integrity.

Anatomy of Liver

The liver is the largest organ in the body weighing 1400-1600 gm in the males and 1200-1400 gm in the females. There are 2 main anatomical lobes right and left lobe, the right being about six times the size of the left lobe. The right lobe has quadrate lobe on its inferior surface and a caudate lobe on the posterior surface. The right and left lobe are separated anteriorly by a fold of peritoneum called the falciform ligament, inferiorly by the fissure for the ligamentum teres, and posteriorly by the fissure for the ligamentum venosum (Figure 1).

The porta hepatis is the region on the inferior surface of the right lobe where blood vessels, lymphatics and common hepatic ducts form the hilum of the liver. A firm smooth layer of connective tissues called Glisson's capsule encloses the liver and is continuous

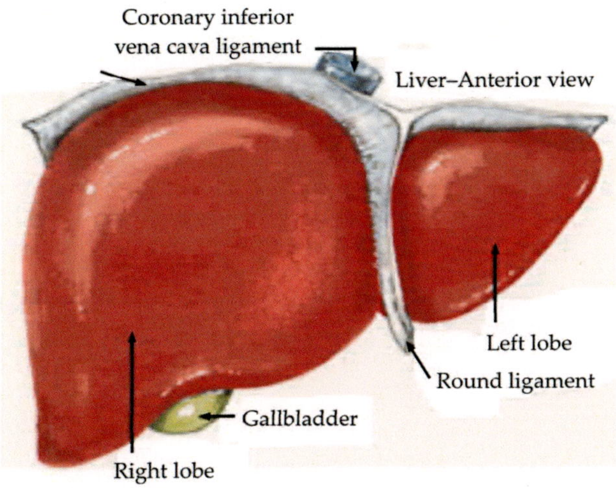

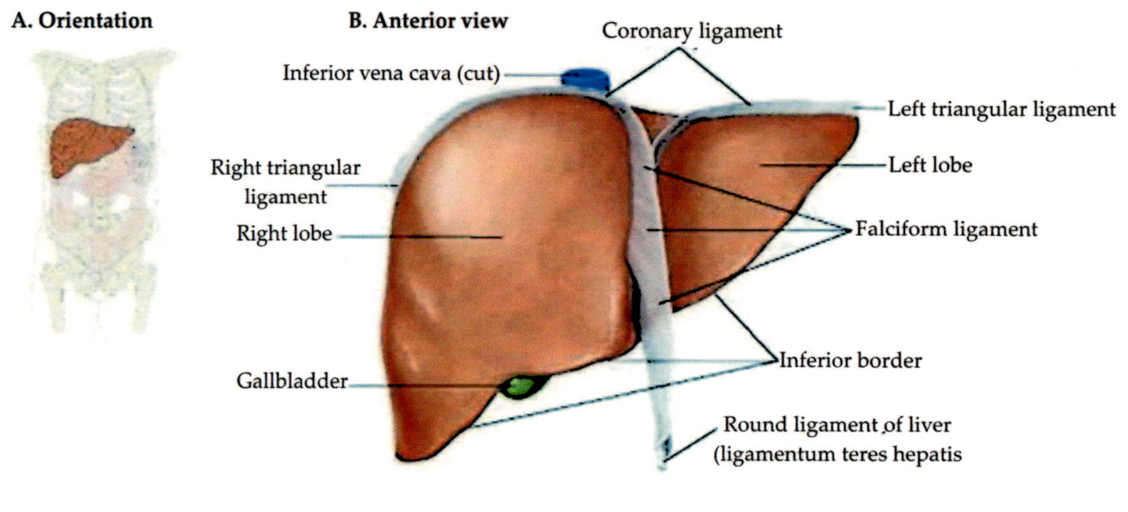

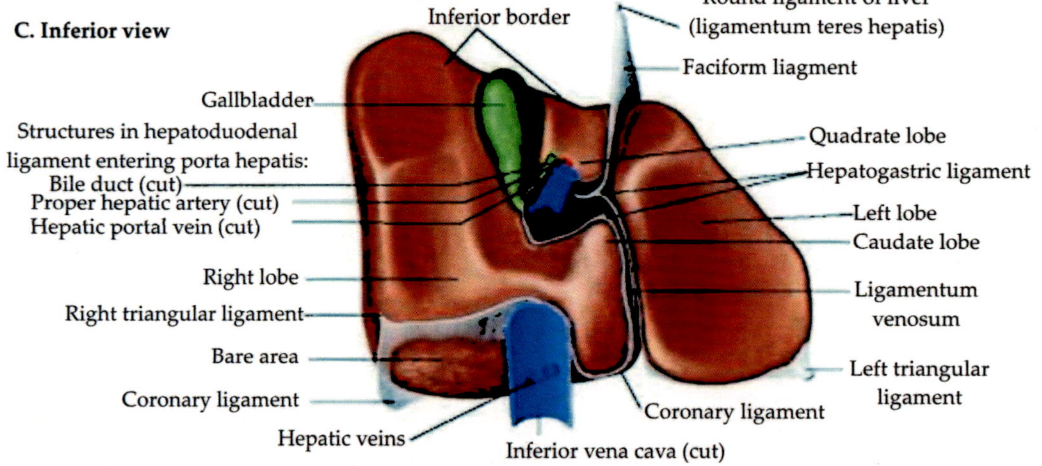

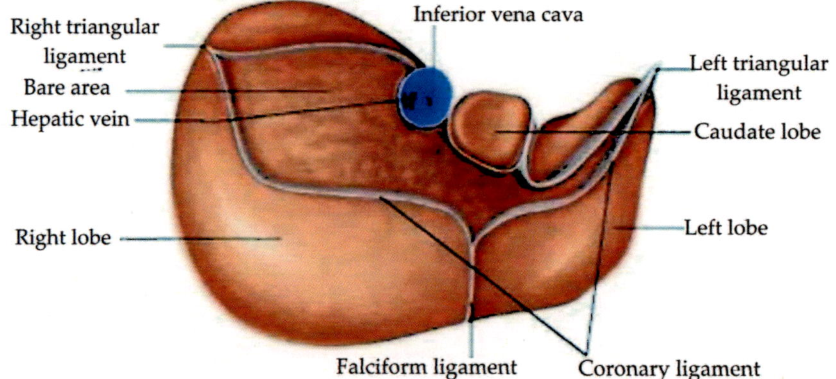

Figure 1: Anatomy of liver

with the connective tissue of the porta hepatis forming a sheath around the structure in the porta hepatis. The liver has a double blood supply-the portal vein brings the venous blood from the intestine and spleen, and the hepatic artery coming from the celiac axis supplies arterial blood to the liver. This dual blood supply provides sufficient protection against infarction in the liver.

The portal vein and hepatic artery divide into branches to the right and left lobes in the porta. The right and right hepatic ducts also join in the porta to form the common hepatic duct. The venous drainage from the liver is into the right and left hepatic veins which enters the inferior vena cava. Lymphatic and the nerve fibers accompanies the hepatic artery into their branching and terminate around the porta hepatic.

Histology of Liver

The hepatic parenchyma is composed of numerous hexagonal or pyramidal classical lobules; each with a diameter of 0.5 to 2 mm. each classical lobule has a central tributary from the hepatic vein and at the periphery are 4 to 5 portal tracks or triads containing branches of bile ducts, portal vein and blood containing sinusoids radiate from the central vein to the peripheral portal triads. The functioning lobules or liver actinus as described by Rapp port has a portal triad in the centre and is surrounded at the periphery by portion of several classical lobules.

The blood supply to the liver parenchyma flows from the portal triads to the central veins. Accordingly, the hepatic parenchyma of liver lobule is divided into 3 zones:

ZONE 1 or the periportal (peripheral) area is closest to the arterial and portal blood supply and hence bears the brunt of all forms of toxic injury.

ZONE 3 or the centrilobular area surrounded the central vein and is most remote from the blood supply and thus suffers from the effects of hypoxic injury.

ZONE 2 is the intermediate midzonal area.

The hepatocytes are polygonal cells with a round single nucleus and a prominent nucleolus. The liver cell have a remarkable capacity to undergo mitosis and regeneration. Thus it is not uncommon to find liver cells containing more than one nuclei and having polyploidy up to octoploidy. A hepatocyte has 3 surface; one facing the sinusoid and space of disse, the second facing the canaliculus, and the third facing neighboring hepatocyte.

The blood-containing sinusoids between cords of hepatocytes are lined by discontinuous endothelial cells and scattered flat kupffer cell belonging to the reticulo-endothelial system.

The portal triad or tract besides containing portal vein radical, the hepatic arteriole and bile ducts, has a few mononuclear cells and a little connective tissues considered to be extension of Glisson's capsule. The portal triads are surrounded by a limiting plate of hepatocytes.

The intrahepatic biliary system begins with the bile canaliculi interposed between the adjacent hepatocytes. The bile canaliculi are simply grooves between the contact surface of the liver cells and are covered by microvilli. These canaliculi into terminal bile ducts or ductules (canal of Hering) which are lined by cuboidal epithelium (Figure 2).

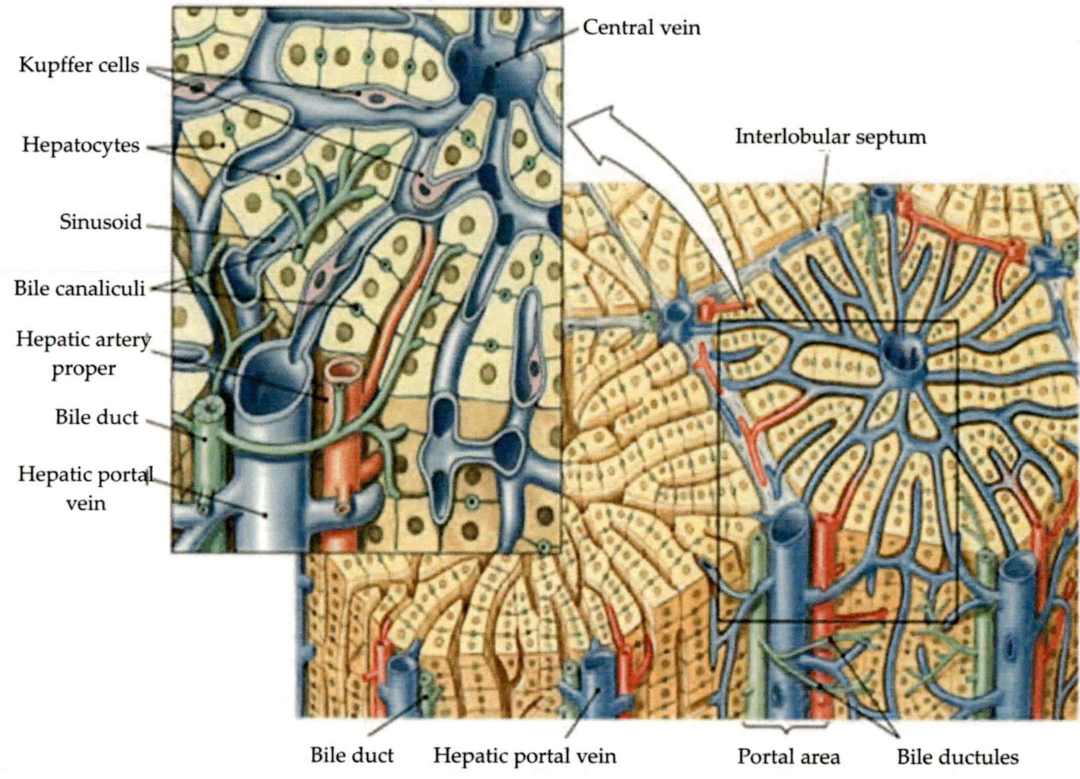

Figure 2: Histology of liver

Functions of Liver

The liver is an extremely active organ. Some of its functions have described here:

1. *Carbohydrate metabolism*: Conversion of glucose to glycogen in the presence of insulin, and converting liver glycogen back to glucose in the presence of glucagon. These changes are important regulators of the blood glucose level. After a meal the blood in the portal vein has a high glucose contents and insulin converts some to glycogen for storage. Glucagon converts this glycogen back to glucose as required, to maintain the blood glucose level within relatively narrow limits.

2. *Fat metabolism*: Desaturation of fat, *i.e.* converts stored fat to a form in which it can be used by the tissue to provide energy.

3. *Protein metabolism:* Deamination of amino acids: removes the nitrogenous portion from the amino acid not required for the formation of new proteins; urea is formed from this nitrogenous portion which is excreted in urine.

   Breaks down genetic material of worn-out cells of the body to form uric acid which is excreted in urine.

Transamination

Removes the nitrogenous portion of amino acid and attaches it to other carbohydrate molecules forming new non-essential amino acid. Synthesis of plasma proteins and most of the blood clotting factors from the available amino acids occurs in the liver.

4. ***Breakdown of erythrocyte and defense against microbes***: This is carried out by phagocytic kupffer cells (hepatic macrophages) in the sinusoids.

5. ***Detoxification of drugs and noxious substances***: This includes ethanol (alcohol) a toxin produced by microbes.

6. ***Metabolism of ethanol***: This follows consumption of alcoholic drinks.

7. ***Inactivation of hormones***: These include insulin, glucagon, cortisol, aldosterone, thyroid and sex hormones.

8. ***Synthesis of vitamin A from carotene***: Carotene is the pro vitamin found in some plants, *e.g.* carrots and green leaves of vegetables.

9. ***Production of heat***: The liver uses a considerable amount of energy has a high metabolic rate and produces a great deal of heat. It is the main heat producing organ of the body.

10. ***Secretion of bile***: The hepatocyte synthesis the constituents of bile from the mixed arterial and venous blood in the sinusoids. These include bile salts, bile pigments and cholesterol.

11. ***Storage***: The substances include-

Fat soluble vitamins: A, D, E, K

Iron, copper

Some water-soluble vitamins, *e.g.* riboflavin, niacin, pyridoxine, folic acid and vitamin B_{12}.

Liver Diseases

Jaundice

This is the yellow pigmentation of the skin, mucus membrane and deeper tissues due to increased bilirubin level in blood. The normal serum bilirubin level is 0.5 to 1.5 mg%. When this exceeds 2 mg%, jaundice occurs.

Types and causes of jaundice

Jaundice is classified into three type's namely haemolytic jaundice, hepatocellular jaundice and obstructive jaundice.

1. *Haemolytic jaundice*

Haemolytic jaundice is otherwise called prehepatic jaundice. During this, the excretory function of liver is normal. But, there is excessive destruction of red blood cells and thus the bilirubin level in blood is increased, the liver cells cannot excrete that much bilirubin rapidly. So, it accumulates in the blood resulting in jaundice.

In this type of jaundice, the free (unconjugated) bilirubin level increases in blood. The formation of urobilinogen also is more resulting in the excretion of more amount of urobilinogen in urine.

Any condition that causes hemolytic anemia can lead to hemolytic jaundice.

2. *Hepatocellular jaundice*

The jaundice due to the damage of liver cells is called hepatocellular or hepatic jaundice.

It is also called hepatic cholestatic jaundice. Here bilirubin is conjugated. But the conjugated bilirubin cannot be excreted. So, it returns to blood. The damage of the liver cells occurs because of toxic substances (toxic jaundice) or by infection (infective jaundice). Commonly liver is affected by virus resulting in hepatitis.

3. *Obstructive jaundice*

This is otherwise called extrahepatic cholestatic jaundice or posthepatic jaundice. It is due to the obstruction of bile flow at any level of the biliary system. The bile cannot be poured into small intestine and bile salts and bile pigments enter the circulation. In this, blood contains more conjugated bilirubin. The important differences between haemolytic jaundice and obstructive jaundice are given in the Table 1.

Hepatitis

Hepatitis is a liver disease characterized by swelling and inadequate functioning of liver. Hepatitis may be acute or chronic. In severe conditions, it may lead to liver failure and death.

Causes and types

Hepatitis is caused by viruses, bacteria, poisons, autoimmune diseases, drug abuse, alcohol, some therapeutic drugs and inheritance from mother during parturition. Viral hepatitis is of five types namely, hepatitis A, B, C, D and E.

Hepatitis A and E are caused mostly by intake of water and food contaminated with hepatitis virus. Generally, these two types of hepatitis are not life-threatening.

Hepatitis B, C and D are caused by sharing needles with infected persons, accidental prick by infected needle, having unprotected sex with infected persons, inheritance from mother during parturition and blood transfusion from infected donors. These three forms of hepatitis are serious diseases when compared to hepatitis A and E. Among these, hepatitis B is more common and considered more serious because it may lead to cirrhosis and cancer of liver.

Cirrhosis

The inflammation and damage of parenchyma of liver is known as cirrhosis of

Table 1: Difference between Haemolytic Jaundice and Obstructive Jaundice

Features	*Haemolytic Jaundice*	*Obstructive Jaundice*
Serum bilirubin level	High with more free bilirubin	High with more conjugated bilirubin
Urobilinogen in urine	More	Absent
Bile salts and pigments in urine	Absent	Present
Van den Bergh's reaction	Indirect-positive	Direct-positive
Liver functions	Normal	Exaggerated
Hemorrhagic tendency	Absent	Present due to lack of vitamin K

liver. This may result in degeneration of hepatic cells and dysfunction of liver.

Cirrhosis is a diffuse, chronic, necrotic (degenerative) liver disorder characterized by progressive hepatocyte injury followed by regeneration and fibrosis leading to disorganisation of lobular architecture, pseudolobule formation and acquired vascular malformation.

Characteristic features

- Disorganised architecture of liver due to necrosis, fibrosis and regeneration.
- Fibrosis-Broad scars.
- Parenchymous nodules (Pseudulobule) formation due to
  - Regenerative activity
  - Fibrosis and scar formation

These nodules are called as

- – Micronodules (if less than 3 mm)
- – Macronodules (if more than 3 mm)
- Development of portal hypertension and their complications.

Complication

Progressive liver failure
Portal hypertension and its complications
Hepatocellular carcinoma.

Tumors of Liver

(A) *Benign tumors*

(i) Benign haemangioma
(ii) Cysts
- Simple cyst
- Conjugated intrahepatic biliary dilatations.
- Choloedochal cyst

Polycystic liver disease.

(i) Focal Hyperplasia
(ii) Adenoma
- Bile duct adenomas
- Liver cell adenomas

(B) *Malignant tumors*

(i) Secondary metastasis are the most common tumors, which may be from breast, lung and colon.
(ii) Primary tumors
- Hepatoblastoma
- Angiosarcoma
- Primary Carcinoma of liver Hepatocellular Carcinoma Cholangiocarcinoma

Hepatocellular Carcinoma

It is the most common primary liver cancers (Comprising 90% of all tumors).

Etiology and Pathogenesis

- Protracted infection with Hepatitis B virus.
- Cirrhosis.
- Environmental toxins *e.g.*, Alfa toxin B; produced by *Aspergillus flavus*.
- Oral contraceptive questionable role.

Hepatocellular Failure

It may occur due to -

(i) Ultra structural lesions of hepatocytes *e.g.*, reye's syndrome, drugs.
(ii) Chronic liver disease *e.g.*, chronic hepatitis, cirrhosis, Wilson's disease.
(iii) Massive hepatic necrosis fulminant viral hepatitis.

Clinical Features

- Jaundice
- Hypoalbuminemia
- Hyperammonemia
- Increased serum levels of hepatic enzymes *e.g.*, SGPT, SGOT.
- Hepatorenal syndrome
- Hepatic encephalopathy
- Coma

Hepatic Encephalopathy

Also called as hepatic coma, is a feature of acute of chronic liver failure. It is a metabolic disorder of the central nervous system and neuromuscular system associated with hepatic failure. It is reversible condition.

Clinical Features

1. Nonspecific, electroencephalographic (EEG) changes.
2. Disturbed consciousness progressing to coma.
3. Fluctuating neurologic signs.
4. Characteristic "flapping tremors" known as asterixis.
5. Progressive confusion, drowsiness and coma which may lead to death.

Hypothesis

- Ammonia hypothesis – less degradation of ammonia to urea by liver leads to toxic damage to brain by hyperammonemia.
- Synergistic neurotoxin hypothesis: encephalopathy due to combined action of ammonia, mercaptans and short chain fatty acids.
- The false neurotransmitter hypothesis: The normal neurotransmitter dopamine in brain is replaced by false neurotransmitter like GABA or octopamine.
- The amino acid hypothesis – *e.g.*, elevated plasma levels of tryptophan and its metabolite, serotonin may be toxic to the brain.

Portal Hypertension

In this condition, increased resistance to portal blood flow. It may occur in the following conditions -

(i) Prehepatic

- Portal vein thrombosis.
- Splenomegaly.

(ii) Intrahepatic

- Cirrhosis
- Miliary Tuberculosis
- Idiopatic.

(iii) Post hepatic

- Severe right heart failure.

Major consequences

(i) Ascites

The collection of excess fluids (>500ml) in peritoneal cavity.

Ascites may occur due to-

- Increased hepatic lymph formation.
- Retention of sodium and water.

(ii) Splenomegaly

It is congestive in nature. It might be upto 1000 gms and may lead to hypersplenism.

(iii) Hepatic encephalopathy

It is usually associated with cirrhosis, portosystemic shunting, sudden drainage of large volume of ascites, massive hematemesis, infection, stress, operation, electrolyte imbalance, drugs or protein loss.

Hepatotoxicity

Many therapeutic drugs cause liver damage, manifested clinically as hepatitis or (in less severe cases) only as laboratory abnormalities (*e.g.,* increased; activity of plasma transaminase, an enzyme released from damaged liver cells). Paracetamol, isoniazid, iproniazid and halothane cause hepatotoxicity by the mechanisms of cells damage outlined above. Genetic differences in drug metabolism have been implicated in some instances (*e.g.,* isoniazid, phenytoin). Mild drug-induced abnormalities of liver function are not uncommon, but the mechanism of liver injury is often uncertain (*e.g.,* statins).

It is not always necessary to discontinue a drug when such mild laboratory abnormalities occur, but the occurrence of irreversible liver disease (cirrhosis) as a result of long-term low-dose methotrexate treatment for arthritis or psoriasis (a chronic scaling skin disease of unknown cause that is usually mild, if tiresome, but can rarely be very severe) argues for caution. Hepatotoxicity of different kind, namely reversible obstructive jaundice, occurs with chlorpromazine and androgens.

General mechanism of cell damage and cell death

Drug-induced cell damage/death is usually caused by reactive metabolites of the drug, involving non-covalent and/or covalent interactions with target molecules. Cell death is often 'self-inflicted', *via* triggering apoptosis.

Non-covalent Interactions Include

- Lipid peroxidation *via* a chain reaction.
- Generation of cytotoxic reactive oxygen species.
- Depletion of reduced glutathione.
- Modification of sulfhydryl groups on key enzymes (*e.g.,* Ca^{2+}-ATPase) and structural proteins.

Covalent interactions, for example adduct formation between a metabolite of Paracetamol (NAPBQI: N-acetyl-p-benzo-quinone imine) and cellular macromolecules. Covalent binding to protein can produce an immunogen; binding to DNA can cause carcinogenesis and teratogenesis.

Hepatotoxins

An agent such as drugs and chemicals which produce toxic liver injury may virtually mimic any form of naturally – occurring liver disease. Hepatoprotective effects were studied against chemicals and drugs induced hepatotoxicity in rats like alcohol, CCl_4, galactosamine, paracetamol, isoniazid, rifampicin etc.

Classification of Hepatotoxins

1. Intrinsic

It consists of agents that are predictable hepatotoxins. They are recognized by high incidence of hepatic injury in exposed individuals and in experimental animals. There is a consistent latent period between exposure to a particular agent and the development of hepatic injury and the injury appeared to be dose related.

There are Two Types of Intrinsic Hepatotoxins:

(a) Direct hepatotoxins

It may be so called because they (or their metabolic products) produce direct injury to hepatocytes and its organals, especially the endoplasmic reticulum. Carbon tetrachloride, the prototype, produces peroxidation of the membrane lipids and other chemicals that lead to degeneration of the membranes.

(b) Indirect hepatotoxins

They are anti-metabolites and related compounds that produce hepatic injury by interference with the specific metabolic pathway or processes. The structural injury produced by indirect hepatotoxins appears to be secondary to a metabolic lesion, while in that produced by direct hepatotoxins, the metabolic derangement is secondary to the structural injury. The hepatic damage produced by indirect hepatotoxin may be mainly cytotoxin injury (by interfering with metabolic pathways or processes essential for parenchyma integrity) expressed as steatosis or necrosis, or may be mainly cholestasis, interfering only or mainly with biliary secretion.

2. *Host idiosyncrasy*

It consists of agents that are not predictably hepatotoxic but produces hepatic injury in only a small proportion of exposed individuals, who are uniquely susceptible. In several instances auto antibodies directed against normal cellular constituents are detected. The injury does not appear to be dose related and is not reproducible in experimental animals and appears after a variable latent period.

The classification of hepatotoxins is shown in Table 2.

Table 2: Classification of hepatotoxins

Category of agent	*Mechanism*	*Histological lesion*	*Examples*
1. Intrinsic toxicity (a) Direct	Membrane injury destruction of structural basis of cell metabolism	Necrosis (zonal) and/or steatosis	CCl_4, $CHCl_3$, phosphorus.
(b) Indirect Cytotoxic	Interference with specific metabolic pathways leads to structural injury	Steatosis or Necrosis thioacetamide, paracetamol, ethanol.	Ethionine,
(c) Cholestatic	Interference with hepatic excretory pathway leads to cholestasis	Bile duct injury	Rifampicin, steroids.
2. Host idiosyncrasy (a) Hypersensitivity (b) Metabolic abnormality	Drug allergy Production of hepatotoxic metabolites	Necrosis or cholestasis Necrosis or cholestasis	Sulphonamides, halothane. Isoniazid

Classification of Hepatotoxicity

Hepatotoxicity

1. *Chemical induced hepatotoxicity*

e.g., Carbon tetrachloride (CCl_4), Ethanol, Chloroform, Allyl alcohol, Arsenic, Diethyl nitrosamine, Dimethylformamide, Aroclor 1254, dimethyl acetamide.

2. Drug induced hepatotoxicity

I. *NSAIDs induced hepatotoxicity*

e.g., Paracetamol , Indomethacin, Aspirin,

II. *Antitubercular drug induced Hepatotoxicity*

e.g., Rifampin, isoniazide, D- Galactosamine.

3. *Phytochemical induced hepatotoxicity*

e.g., Amanitin(*Amanita phalloides*), Phalloidin (*Amanita phalloides*), Aflatoxin (*Aspergillus flavus*), Luteoskyrin (*Penicillium islandicum*), Cycasin (*Cycas circinalis* L.).

1. *Carbon tetrachloride (CCl_4) Induced Hepatotoxicity*

Delayed toxic effects of $CC1_4$ include nausea, vomiting, abdominal pain, diarrhea and hematemesis. The most serious delayed toxic effect of CCl_4 results from its hepatotoxic and nephrotoxic actions. Signs and symptoms of hepatic injury may appear after a delay of several hours or 2 to 3 days and may occur in the absence of earlier severe effects on the CNS. Biochemical evidence of hepatic injury often includes greatly elevated activities of transaminases and a variety of other hepatic enzymes in plasma. Alkaline phosphatase activity is, however, only slightly elevated. The chief histological abnormalities include hepatic steatosis and hepatic centrilobular necrosis.

Mechanism

The mechanism of $CC1_4$-induced hepatic injury has interested many investigators and the compound has become the reference substance for all hepatotoxic compounds. Injury produced by $CC1_4$ seems to be mediated by a reactive metabolite-trichloromethyl free radical ($.CC1_3$) formed by the hemolytic cleavage of $CC1_4$, or by an even more reactive species-trichloromethylperoxy free radical ($Cl_3COO.$)-formed by the reaction of $CC1_3$ with O_2. This biotransformation is catalyzed by a cytochrome P450-dependent monooxygenase. Thus, agents such as DDT and Phenobarbital, which induce such enzymes, strikingly enhance the hepatotoxic effects of $CC1_4$. Conversely, agents that inhibit the drug-metabolizing activity diminish the hepatotoxicity of $CC1_4$. Biotransformation of $CC1_4$ to the reactive intermediate is irreductive rather than an oxidative reaction. As a result, it is slower at high oxygen tensions.

The toxicity produced by $CC1_4$ is thought to be due to the reaction of free radicals ($.CC1_3$ or $C1_3COO.$) with lipids and proteins; however, the relative importance of interactions with various tissue constituents in producing injury is controversial. The free radical causes the peroxidation of the polyenoic lipids of the endoplasmic reticulum and the generation of secondary free radicals derived from these lipids-a chain reaction. This destructive lipid peroxidation leads to breakdown of membrane structure and function, and, if a sufficient quantity of CCl_4

has been consumed, the intracellular cytoplasmic Ca^{2+} increases, resulting in cell death. Lipid peroxidation leads to a unique series of nonenzymatic arachidonic acid metabolites, dubbed *isoprostanes* in parallel with CCl_4-induced hepatotoxicity in laboratory animals; these agents may serve a diagnostic role in identification of lipid peroxidation in human beings. Kupffer cells also participate in a mechanism of carbon tetrachloride toxicity, probably by releasing chemoattractants for neutrophils that produce more oxidative stress.

2. Ethanol Induced Hepatotoxicity

Ethanol produces a constellation of dose-related deleterious effects in the liver. The primary effects are fatty infiltration of the liver, hepatitis, and cirrhosis. Because of its intrinsic toxicity, alcohol can injure the liver in the absence of dietary deficiencies. The accumulation of fat in the liver is an early event and can occur in normal individuals after the ingestion of relatively small amount of ethanol. This accumulation results from inhibition of both the tricarboxylic acid cycle and the oxidation of fat, in part owing to the generation of excess NADH produced by the actions of alcohol dehydrogenase and aldehyde dehydrogenase.

Fibrosis, resulting from tissue necrosis and chronic inflammation, is the underlying cause of alcoholic cirrhosis. Normal liver is replaced by fibrous tissue. Alcohol can affect directly stellato cells in the liver, causing deposition of collagen around terminal hepatic venules (Worner and Lieber, 1985). Chronic alcohol use in associated with transformation of stellate cells into collagen producing, myofibroblast-like cells (Lieber, 1998). The histologic hallmark of alcoholic cirrhosis is the formation of Mallory bodies, which are thought to be related to an altered cytokeratin intermediate cytoskeleton. A number of underlying molecular mechanisms have been proposed.

3. Paracetamol Induced Hepatotoxicity

In adults, hepatotoxicity may occur after ingestion of a single dose of 10 to 15 g (150 to 250 mg/kg) of acetaminophen; does not 20 to 25 g or more are potentially fatal. Alcoholics can have hepatotoxicity with much lower doses, even with doses in the therapeutic range. Symptoms that occur during the first 2 days of acute poisoning by acetaminophen may not reflect the potential seriousness of the intoxication. Nausea, vomiting, anorexia, diaphoresis, and abdominal pain occur during the initial 24 hours and may persist for a week or more. Clinical indications of hepatic damages become manifest within 2 or 4 days of ingestion of toxic doses. Plasma amino transferases are elevated (sometimes markedly so), and the concentration of bilirubin in plasma may be increase; in addition, the prothrombin time is prolonged. Perhaps 10% of poisoned patients who do not receive specific treatment develop sever liver damage; of these, 10% to 20% eventually die of hepatic failure. Acute renal failure also occurs in some patients. Biopsy of the liver reveals centrilobular necrosis with sparing of periportal area. In nonfatal cases, the hepatic lesions are reversible over a period of weeks or month.

Severe liver damage (with levels of aspartate aminotransferase activity in excess of 1000 IU per litre of plasma) occurs in 90 % of patients with plasma concentration of acetaminophen greater than 300 μg/ml at 4 hours or 45μg/ml at 15 hours after the ingestion of the drug. Minimal hepatic damage can be anticipated when the drug concentration is less than 120 μg/ml at 4 hours or 30 μg/ml at 12 hours after ingestion.

The potential severity of hepatic necrosis also can be predicted from the half-life of acetaminophen observed in the patient; values greater than 4 hours imply that necrosis will occur, while values greater than 12 hours suggest that hepatic come is likely.

Mechanism

The metabolites of many chemicals are responsible for their toxicities. Most organophosphate insecticides are biotransformed by the cytochrome P 450 system to produce their toxicities; for example, parathion is biotransformed to paraoxon. Paraoxon is a stable metabolite that binds to and inactivates cholinesterase. Some metabolites of drugs are not chemically stable and are referred as reactive intermediates. An example of a toxic reactive intermediate is the metabolite of acetaminophen, which is very reactive and binds to nucleophiles such as glutathione; when cellular glutathione is depleted, the metabolite binds to cellular macromolecules, the mechanism by which acetaminophen kills liver cells (Figure 3).

Both parathion and acetaminophen are more toxic under conditions in which the cytochrome P450 enzymes are increased, such as following ethanol or Phenobarbital exposure, because there are responsible for production of the toxic metabolites.

4. Isoniazid Induced Hepatotoxicity

The incidence of adverse reactions to isoniazid was estimated to be 5.4% among more than 2000 patients treated with the drug; the most

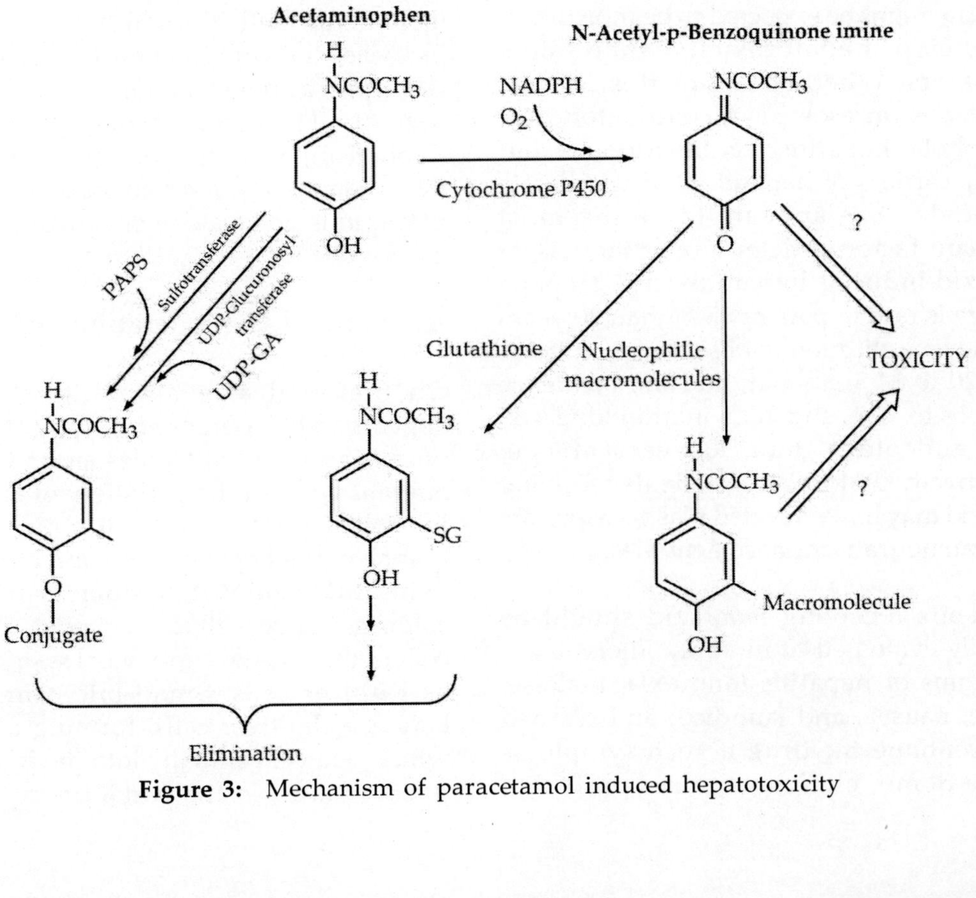

Figure 3: Mechanism of paracetamol induced hepatotoxicity

prominent of these reactions were rash (2%), fever (1.2%), jaundice (0.6%) and peripheral neauritis (0.2%).

Jaundice has been known for some time to be an untoward effect of exposure to isoniazid, not until the early 1970s did it become apparent that severe hepatic injury leading to death may occur in some individuals receiving this drug. Additional studies in adults and children have confirmed this observation; the characteristics pathological process is bridging and multilabular necrosis. Continuation of drug after symptoms of hepatic dysfunction have appeared tends to increase the severity of damage. The mechanisms responsible for this toxicity are unknown, although acetylhydrazine, which is a metabolite of isoniazid, causes hepatic damages in adults. Hence, patients who are rapid acetylators of isoniazid might be expected to be more likely to develop hepatotoxicity than slow acetylators. Whether or not this is true, however, is unresolved. A contributory role of alcoholic hepatitis has been noted, but chronic carriers of hepatitis B virus tolerate isoniazide. Age appears to be the most important factor in determining the risk of isoniazid-induced hepatotoxicity. Hepatic damage is rare in patients less than 20 years old; the complication is observed in 0.3% of those 20 to 34 years old, and the incidence increases to 1.2% and 2.3% in individuals 35 to 49 and older than 50 years of age, respectively. Up to 12% of patients receiving isoniazid may have elevated plasma aspartate and alanine transaminase activities.

Patients receiving isoniazid should be carefully evaluated at monthly intervals for symptoms of hepatitis (anorexia, malaise, fatigue, nausea, and jaundice) and warned to discontinue the drug if such symptoms occur. Some clinicians also prefer to determine serum aspartate amino transferase activities at monthly intervals and recommended that an elevation greater than five times normal is cause for discontinuation of the drug. Most hepatitis occurs 4 to 8 weeks after the start of therapy. Isoniazid should be administered with great care to those with preexisting hepatic disease.

5. Rifampin Induced Hepatotoxicity

Rifampin generally is well tolerated. When given in usual doses, fewer than 4% of patients were tuberculosis have significant adverse reactions; the most common are rash (0.8%), fever (0.5%) and nausea and vomiting (1.5%). Rarely, hepatitis and death due to liver failure have been observed in patients who received other hepatoxic agents in addition to rifampin or who had preexisting liver disease. Hepatitis from rifampin rarely occurs in patients with normal hepatic function; likewise, the combination of isoniazid and rifampin appears generally safe in such patients. However, chronic liver disease, alcoholism, and old age appear to increase the incidence of severe hepatic problems when rifampin is given alone or concurrently with isoniazid.

Anatomical Considerations of Liver

On histological examination, the liver appears to be composed of radical columns of cells arranged in lobules around a central efferent vein. At the periphery of the lobules lie portal tracts, each containing a small branch of the hepatic artery and of the portal vein and a small bile duct. Between the columns of liver cells are sinusoids lined with cells of the reticulo endothelial system known as Kupffer cells. Small bile canaliculi lie between the liver cells, forming a network which opens eventually into the interlobular ducts situated in the portal tracts.

From the portal vein and from the hepatic artery in the portal tract, blood passes into the sinusoids and reaches the central vein, which drains into the hepatic veins. Mixing of portal venous and hepatic arterial blood appears to occur in the sinusoids. When the normal architecture of the liver, including presinusoidal sphincters on the arterioles is damaged by disease, direct transmission of the high arterial pressure to the portal system may be partly responsible of the portal venous hypertension found in such conditions.

Bile, secreted by the liver cells, passes, in reverse direction of the blood flow, through canaliculi to the periphery of the lobule. There, the bile ducts are lined with cuboidal epithelium and gradually become larger as they progress to the porta hepatic. Ultimately the common hepatic duct is formed by the union of the ducts from the right and left lobes of the liver, and this, with the cystic duct form the right and left lobes of the liver, and this, with the cystic duct from the gall-bladder, forms the common bile duct. This usually traverses the substance of the head of the pancreas, where it is joined, in the majority of cases, by the pancreatic duct. The united ducts then open into the second part of the duodenum through the ampulla of Vater, the opening being controlled by the sphincter of Oddi.

Blood Supply

The liver is unique in that the greater part of the blood flowing to it is venous and comes *via* the portal vein which drains the large and small intestine, stomach, pancreas and spleen. The total blood flow through the adult human liver averages 1,500 m/minute. This frequently becomes considerably reduced in chronic hepatic disease. Though the hepatic artery supplies only about 20 per cent of the blood to the liver, it carries up to 50 percent of the oxygen utilized by the organ. The large supply of venous blood results in an oxygen supply that is probably always precarious, so that the liver is vulnerable to chronic hypoxia. In chronic passive congestion, for example, there is degeneration and sometimes necrosis of the cells furthest from the entry of well-aerated blood, *i.e.* in the centre of lobule. Centrilabular necrosis is found also in thyrotoxicosis, in which the oxygen demand by the liver cells is increased. In addition, in the majority of acute toxic of infective liver diseases, in which there is neither hypoxia nor excessive oxygen utilization, necrosis in this zone is commonly found. Damage to the parenchymal cells in these conditions may be aggravated by the relative oxygen deficiency in the centre of the lobule.

The blood supply to the liver through the portal vein is believed to follow two main streams, one from the stomach, spleen and descending colon to the left, and other from the small intestine and ascending colon to the right lobe. This distribution of blood does not correspond strictly to the anatomical lobes of the liver. Acute massive necrosis, which may in part, at least, be due to dietary inadequacy, may show damage predominantly in the left lobe, presumably because essential dietary constituents are especially deficient in blood draining the stomach, spleen, and descending colon. Like wise, abscesses from an infected appendix occur mainly in the right lobe.

Metabolic Activities of the Liver

The manner in which blood gains intimate contact with hepatic cells facilitates the rapid transfer of metabolites. Apart from the Kupffer cells of the reticuloendothelial system, all the cells of the liver are undifferentiated and appear capable of performing the many functions of the liver. It plays an important role in the maintenance of internal environment through its multiple diverse functions. It involved in the intermediatory metabolism of proteins, fats and carbohydrates and in the synthesis of number of plasma proteins, such as albumin, fibrinogen and clotting factors in the precautions of varied enzymes and formation and excretion of bile.

1. *Carbohydrate metabolism*

The liver is the important organ in the body for the maintenance of a normal concentration of blood glucose. It is able to convert glucose, fructose, galactose, glycerol, certain amino acid residues, and 2-and 3-carbon compound (*e.g.*, lactate, pyruvate and oxaloacetate) to glycogen. In hypoglycaemia it hydrolyses stored glycogen to glucose. These mechanisms are intrinsic function of the organ but they are influenced by many extra hepatic factors, *e.g.*, insulin, adrenaline, thyroxine, cortisol and glucagon.

2. *Protein metabolism*

The liver is the most important site of deamination of amino acids. As a preliminary step in their interconversion and oxidation. Urea synthesis from the amino groups made available by this process occurs solely in the organ. Plasma albumin, prothrombin, fibrinogen and the other clotting factors, V, VIII, IX and X are synthesized perhaps exclusively in the parenchymal cells of the liver. Alfa and beta globulins are mainly formed in the liver, while gama globulin is formed by cells of the reticuloendothelial system. As a result of this synthesis, the protein pattern of the plasma is to a large extent determined by liver function.

3. *Lipid metabolism*

The liver plays an important part in fat metabolisms. Fats are oxidized in the liver as far as the four carbon chain stage (ketone bodies). The amount of this oxidation depends, inter alia, upon the availability of carbohydrate and is greatly increased when sugar is absent as in fasting or when there is difficulty in its use as in diabetes mellitus. The ketone bodies themselves do not appear to be oxidation. Triglycerides are formed in the liver and synthesis of phospholipid from fatty acid, glycerol, phosphate and a nitrogenous base, *e.g.*, choline, also occurs largely in the liver. Cholesterol is synthesized in the liver and blood as lipoproteins; not only is the liver an important source of these complexes which maintain lipid in colloidal solution, but also in colloidal solution, but also it contains an enzyme which can break the lipid-protein bond. Bile salts, the breakdown products of cholesterol, are synthesized in the liver cells and in turn facilitate the excretion of cholesterol in the bile.

4. *Vitamin metabolism*

The liver is directly or indirectly concerned with the metabolism of many vitamins. Fat-soluble vitamins depend to some extent for their absorption upon a normal biliary secretion. Vitamin K is required by the hepatic cells for the production of prothrombin and factor VII. The liver contains enzymes and prosthetic groups. The phosphorylation of thiamine, which is a necessary preliminary to its function as a coenzyme, the methylation of nicotinic acid and the 25-hydroxylation of vitamin D also occur in the liver.

5. *Inactivation of hormones*

Oestrogens, cortico steroids and other steroid hormones are conjugated in the liver with glucuronic acid and excreted in the urine, though other modes of inactivation also occur. Thyroxine and vasopressin are probably inactivated in the liver, but the mechanisms are unknown.

6. *Detoxification of drugs*

The liver plays a vital role in the detoxification of drugs as well as endogenous hormones. This is achieved by enzymes systems of broad specificity located on the endoplasmic reticulum (microsomal enzymes). At these sites the drugs are rendered more water soluble and can then be excreted in the bile and urine. Alkaloids such as morphine or atropine are partly destroyed in the liver, ammonia is converted to urea, and barbiturates undergo oxidation of their side chain and are rendered pharmacologically inert. Conjugation with glucuronic acid occurs with salicylates, morphine and chloral hydrate, and these conjugates are less active than their precursors. Acetylation of sulphonamides also occurs though this process makes these drugs less soluble and potentially more harmful.

The duration and intensity of action of many drugs are largely determined by the speed at which they are metabolized by these microsomal enzymes of the liver. The activity of the enzymes can be altered by dietary and nutritional factors or alterations in hormonal balance. The activities of the enzymes may also be markedly increased by the coincident administration of drugs. This increase in activity appears to represent an increased concentration of enzyme protein; it is referred to as enzyme induction and may be of great significance in man because of the consequent alteration in duration and intensity of drug action. This enzymes induction may have important therapeutic implications, *e.g.*, a decreased response to coumarone anticoagulants whilst taking barbiturates followed by a markedly diminished dose requirement of withdrawal of the barbiturate. Furthermore, some drugs can stimulate their own metabolism, an effect which may explain the increased tolerance for alcoholic.

7. *Biliary excretion*

Bile is produced by the liver and passes through the intrahepatic biliary channels to the common hepatic duct and then to the gall-bladder. The main constituents, apart from water and inorganic salts, are bilirubin, bile acids, cholesterol, alkaline phosphatase and mucin.

(a) Bilirubin

Hemoglobin is broken down by reticulo-endothelial cells mainly in the spleen, liver and bone marrow. The bile pigment, bilirubin, is derived from the non-iron-containing residue of hemoglobin, after the separation from globin. A small amount of bilirubin is derived from other haem-containing compounds such as hemoglobin precursors in the bone marrow, myoglobin and the cytochromes. In the blood this unconjugated bilirubin (prehepatic bilirubin) is bound to plasma albumin and, as a consequence, does not pass readily through the glomerulus of the kidney into the urine. Unconjugated bilirubin is actively its passage it is separated from its protein and is conjugated with glucuronic acid. Conjugation occurs in the microsomal of the liver cells by the activity of glucuronyl transferase and renders the pigment water-soluble so that it becomes capable of being more freely excreted in the urine. This compound is called conjugated bilirubin. In patients with unconjugated hyper bilirubinemia the level of bilirubin in the plasma can be lowered by the administration of barbiturates which stimulate microsomal enzyme activity.

In the small intestine conjugated bilirubin is metabolized by bacteria to a series of isomers of stercobilinogen which on oxidation form the main faecal pigment stercobilin. The bulk of these pigments is excreted in the stool (250 mg/day) but some is reabsorbed from the gut. Most of this is excreted by the liver cells into the intestine and a small part is excreted in the urine (2-4 mg/day) where it is called urobilinogen. Urobilinogen and its oxidation product urobilin are chemically identical with stercobilinogen and stercobilin, respectively.

(b) Excretion of Bile Acids and Cholesterol

Bile acids are the breakdown products of cholesterol. They are synthesized in the liver and conjugated with glycine and taurine to from bile salts which are secreted directly into the bile. Here they may form micelles when in sufficient concentration and these micelles, which also contain phospholipid, permit the solubilisation of cholesterol although its biliary concentration far exceeds its aqueous solubility in a simple solution. An alteration of these solutes in bile is now recognized as a major factor in the precipitation of cholesterol and provided the nidus for gall-stone formation.

In the lumen of the proximal small intestine bile salts are essential for the solubilisation of the products of fat digestion. Lack of this detergent property of bile explains the steatorrhea which, although often latent, is a well recognized feature of hepatobiliary disease. In addition to dietary glycerides there can be malabsorption of fat-soluble vitamins leading to a bleeding tendency (Vitamin k_0, osteomalacia (vitamin D) and night blindness (vitamin A). About 90 percent of the bile salts are reabsorbed from the distal ileum and return to be re-excreted by the liver. This very efficient enterohepatic circulation preserves the small total body pool of bile salts which may recycle two to three times per meal. Only a small amount (300 mg) of bile salts in the faeces. In the colon the bile salts have a cathartic action due interference with the resorption of electrolytes and water. In addition to a troublesome diarrhea there is also steatorrhea since the liver cannot compensate for the duodenum is insufficient to form micelles. In patients with an over growth of bacteria in the small bowel (blind loop syndrome), there may be malabsorption of lipid is believed to be the result of bacterial alteration of the bile salt molecule (deconjugation and dehydroxylation) so that the effective concentration of bile structure of bile salts may also occur in patients with cholangitis. Whilst in pruritus there may be elevated levels of bile slats in the skin and plasma a definite relationship between bile salts and prutitus has not yet established.

(c) Excretion of enzymes

The serum alkaline phosphatase is derived mainly from the intrahepatic biliary system and the level tends to rise when there is obstruction to biliary excretion. An isoenzyme produced by the osteoblasts accounts for a variable proportion of alkaline phosphatase activity in growing children and in disease of bone. In about 50 per cent of healthy individuals there is in the serum a small amount of alkaline phosphatase which is produced by the intestinal mucosa. Yet another isoenzyme is produced by the placenta and accounts for some of the serum alkaline phosphatase activity in pregnancy. Various malignant tumours, *e.g.,* of lungs and pancreas, have been shown to produce alkaline phosphatase which has similar physical-chemical properties to the placental isoenzyme. The different isoenzymes can be identified by starch gel electrophoresis and studies of enzyme susceptibility to heat inactivation.

Hence, any injury to liver or impairments of its function has grave implication for the health of the affected person. Every year about 18000 people are reported to die to liver cirrhosis causes by hepatitis. Although viral infection is one of the main causes of hepatic injury, xenobiotics, excessive drug therapy, environmental pollutants and chronic alcohol ingestion can also cause severe liver injury since it plays a central role in processing, metabolizing and disposition of foreign chemicals it is susceptible to their injurious effect.

Though liver diseases are among the important diseases affecting mankind, no remedy is available to majority of them at present. However, number of medicinal preparations have been advocated in traditional systems of medicine, especially in Ayurveda, for treating liver disorder. Their usage is in vogue since centuries and are quite often claimed to offer significant relief. In addition usage of many folkore remedies, mainly plant products, is also quite common throughout India. Inspite of such widespread use interest in hepatoprotective activity kindled, especially outside India, only after the publication of the report on isolation of silymarin, a flavonolignan from silybum marianum and its attention of research workers throughout the world towards medicinal plants to search for hepatoprotective agents among them. The search was further facilitated by the description of *in vitro* technique involving evaluating test drugs against carbon tetrachloride induced cytotoxicity in cultured hepatocyte by Hikino and coworkers which enables the research workers to screen large number of test drugs by adopting this procedure as a primary screen.

There is paucity of reviews on medicinal plants possessing hepatoprotective activity. Hence it was thought worthwhile to collect and enumerate data on medicinal plants possessing hepatoprotective activity so that it could serve as a source of information to provide an idea about the current trends in research on plants possessing hepatoprotective activity.

PLANTS FOR HEPATOPROTECTIVE

1. *Acacia catechu* Willd. (Mimosaceae)

Common names

Eng.: Black catechu
Hindi: Katha
Beng.: Khair

Distribution

A small tree widely distributed in drier regions of India, Punjab, M.P., U.P, Bihar, A.P, Orissa and Rajasthan.

Parts used: Heart wood of 20-30 yrs. old plants.

Pharmacological activities

Hepatoprotective activity against CCI, induced hepatic injury was found.

Chemical constituents

Cyanidanol (+) has been isolated from the plant.

2. *Allium sativum* L. (Liliaceae)

Common names
Eng.: Garlic
Hindi: Lahsan
Beng.: Rasun

Distribution

Native to central India, grown in several places in India.

Parts used: Bulbs.

Pharmacological activities

Mechanism of action of hepatoprotection has been reported. The hepatoprotective activity is due to antioxidative property of S-allylmercaptocysteine and S-methylmercaptocysteine inhibiting CCI_4 induced free radical formation measured by the electron spinresonance (ESR) and lipid peroxidation.

Chemical constituents

See Anti-inflammatory Chapter.

3. *Andrographis paniculata* Wall. (Acanthaceae)

Common names
Eng.: King of bitters
Hindi: Kiryat, Kalmegh
Beng.: Kalmegh

Distribution

A branched annual herb occurring throughout India mainly in the plains.

Parts used: Leaves.

Pharmacological activities

The active compound andrographolide and neoandrographolide possess hepatoprotective activity which has been reported to be effective in preventive liver damage caused by *Plasmodium berghei.*

Chemical constituents

Andrographolide has been isolated from the plant. Stereostructure of a diterpene glucoside neoandrographolide was also isolated. Andrographin, panicolin, apigenin 4,7-dimethyl ether and mono-O-methyl within were isolated from roots. Caffeic, chlorogenic and dicaffeylquinic acid were isolated from leaves.

4. *Artemisia capillaris* Linn. (Asteraceae)

Distribution

North-East Himalaya.

Parts used: Bulbs.

Pharmacological activities

The active compounds from bulb of the plant studied against CCI_4 and Gal-N induced cytotoxicity in primary cultured hepatocytes.

Chemical constituents

Capillaries, axcapillin, quercetin and isorhamnetin isolated from buds.

5. *Baccharis trimera* Linn. (Asteraceae)

Parts used: Whole plant.

Pharmacological activities

The active compound phalloidin isolated from the plant possesses hepatoprotective activity (unpublished work).

Chemical constituents

Phalloidin, a flavonoid compound has been isolated.

6. *Boerhaavia diffusa* L. (Nyctaginaceae)

Common names
Eng.: Horse-Purslane, Hogweed
Hindi: Punarnave
Beng.: Punarnave

Distribution

A herb distributed throughout India.

Parts used: Roots and aerial plants.

Pharmacological activities

Chloroform and methanolic extracts of the root and aerial parts of the plant showed hepatoprotective activity against CCI_4 induced liver injury.

Chemical constituents

The active compounds retinoid, steroid, a flavone isolated from chromatographic fractionation. Another active principle punarnavin has been isolated. Hentriacontane, B-sitosterol and ursolic acid isolated from root; a polysaccharide isolated which on hydrolysis yielded glucose, galactose, α-arabinose and α-rhamnose, a glycol-protein with molecular weight of 16000-20000 deltons isolated from roots.

7. *Boerhaavia repanda* Linn. (Nyctaginaceae)

Distribution

Himalayan tracts of Bengal.

Parts used: Roots.

Pharmacological activities

Extracts of root of the plant responded hepatoprotective activity in rats against hepatotoxicity induced by CCI_4, galactosamine (Gal-N) and paracetamol. Petroleum ether, chloroform and methanolic extracts showed significant activity against CCI_4 induced liver injury chloroform and methanolic extracts showed response against Gal-N induced hepatotoxicity. Both the extracts failed to inhibit paracetamol induced liver injury. None of the extracts of the aerial parts exhibited any hepatoprotective activity against CCI_4 induced liver damage.

8. *Butea monosperma* Lam. (Papilionaceae)

Common names
Eng.: Flame of the forest, Kino, Bengal
Hindi: Dhak
Beng.: Palas

Distribution

A deciduous tree occurring throughout India. Bears bright orange red flowers.

Parts used: Fresh juice and gum from the stem.

Pharmacological activities

Hepatoprotective effect has been reported in isobutrin (3,4,2,4-tetrahydroxychalcone 3,4-

diglucoside), butrin (7,3,4-trihydroxy flavanone, 7,3-diglucoside) and two flavonoids isolated from flower extract of the other plant along with the other studies in 20(S)-ginsenoside-RH22$_2$-ginsenoside RG$_3$ and 20(S)-ginsenoside-RS against CCI_4 included cytotoxity.

Chemical constituents

Isobutrin (3,4,2,4-tetrahydroxychalcone 3,4-diglucoside), butrin (7,3,4-triydroxy flavanone, 7,3-diglucoside) and two flavonoids have been isolated from seeds; two new acidic compound along with butrin, Isobutrin, coreopsis, isocoreopin, and sulfurein and structures of monospermoside and isocoreopsin and sulfurein and structures of monospermoside determined isolation and structure of laccijalaric ester I, II and laccijalaric ester III, IV form soft resin α-amyrin, B-sitosterol, its glucoside sucrose isolated from seeds; isolation and structure of a new lactone n-heneicosanoic acid-d-lactone m.p. 70^0 from seeds.

9. *Calotropis gigantea* Linn. (Asclepiadaceae)

Common names
Eng.: Madar
Hindi: Ak, Lal madar
Beng.: Akanda

Distribution

A perennial under shrub distributed throughout India. It is an erect perennial shrub growing chiefly in waste lands. It ascends to an altitude of 3000 ft on Himalayas', extend from Punjab to south India, Assam, Ceylon and Singapore and is distributed to Malaya islands and South China.

Parts used: Flowers.

Pharmacological activities

Hepatoprotective activity against CCI_4 induced liver damage has been found in the ethanolic extracts of the flowers of the plant.

Chemical constituents

The latex which is present in all part of the plant contains water soluble matter 86.0-95.5% and caoutchoue 0.6-1.0%. The coagulum consists of caoutchauc 5.1-18.6%, resins 73.6-87.8% and insoluble matters 4.5-13.8%. Latex contains α-calotropeol ($C_{30}H_{50}O$,m.p.204.5^0) and β-calotropeol (m.p. 216-217^0) mainly in ester combination with acetic and isocaleric acid and B-amyrin.

10. *Canarium album* Raeusch. (Burseraceae)

Pharmacological activities

Brevifolin, hyperin, ellagic acid and 3'-di-O-methyl-ellagic acid isolated from the plant were found to be an effective hepatoprotective compound which has been ascribed due to their antioxidative effects.

Chemical Constituents

Brevifolin, hyperin, ellagic acid and 3'-di-O-methyl-ellagic acid have been isolated from the plant.

11. *Canarium bengalense* Roxb. (Burseraceae)

Common names
Eng.: East Indian copal
Assam: Nerebi
Sylhet: Dhuna

Distribution

A tree occurring in Assam and North Bengal.

Parts used: Leaves.

Pharmacological activities

Leaves possesss hepatoprotective activity.

12. *Canscora decussata* Schult. (Gentianaceae)

Common names
Hindi: Shankhaphuli
Beng.: Dankuni
Sans.: Sankhapushpi

Distribution

A herb occurring throughout India.

Parts used: Whole plants.

Pharmacological activities

Mangiferin isolated from the plant protected rats against CCI_4 induced liver injury.

Chemical constituents

Mangiferin, a xanthone has been isolated from the plant. Gluanone, Canscoradione, friedelin, friedelan-3-Beta-ol, Beta-amyrin, sitosterol, stigmasterol and campesterol was isolated from aerial parts.

13. *Cichorium intybus* Linn. (Asteraceae)

Common names
Eng.: Chicory, Wild endive
Hindi: Kasni

Distribution

The chicory is native to Europe, found in India, cultivated at elsewhere.

Parts used: Roots.

Pharmacological activities

Dried root is used in homoeopathy for liver discuss.

Chemical constituents

The plants contain quercetrin, apigenin, hyperfine, luteolin,-7-O-L-arabinoside, chlorogen and neochlorogenic and inflorescence contains umbelliferone, 6,7-dihydroxy coumarin and esculin.

14. *Curcuma domestica* Val. (Zingiberaceae)

Syn. **C. longa** Linn.

Common name
Eng.: Turmeric
Hindi: Haldi
Beng.: Halud

Distribution

A perennial herb cultivated mainly in Tamil Nadu, Andhra Pradesh, Maharashtra, Bihar, Kerala and Orissa.

Parts used: Rhizomes.

Pharmacological activities

Hepatoprotective activity against CCI_4 induced liver damage has been observed in ethanolic extract of the rhizome of the plant.

Chemical constituents

See Antifertility Chapter.

15. *Cynara scolymus* Linn. (Asteraceae)

Common names

Eng.: Globe
Hindi: Hathichak
Beng.: Hathichoke

Distribution

Bihar and W. Bengal.

Parts used: Whole plant.

Pharmacological activities

Hepatoprotective activities in Cynarine, a polyphenolic compound against CCI_4 and Gal-N induced cytotoxicity in cultured rat hepatocyts have been observed.

Chemical constituents

Cynarine, a polyphenolic compound has been isolated from Italian plant; cynaropicrin and dehydrocynaropicrin isolated from polish plant; absolute stereochemistry or cynaropicrin, dehydrocynaropicrin and grosheimin; hydroxymethylacrylic acid isolated from leaves. Cynaropicrin isolated from leaves and structure elucidated; taraxasterol, Beta-sitosterol, stigmasterol and cynarogenin from receptacles; cynarin, m.p. 2350, from leaves, caffeic acid, 1-,3-,4-and 5-caffeoyl 1,4-dicaffeoylquinic acids, luteolin-7-Beta-D-glucoside and 7-Beta-rutinoside from leaves.

16. *Dianthus superbus var calamus* Linn. (Caryophyllaceae)

Pharmacological activities

D-lanoside H, A, F, E, and D isolated from the plant showed hepatoprotective activity.

Chemical constituents

D-lanoside H, A, F, E and D isolated from the plant.

17. *Emblica officinalis* Gaertn. (Euphorbiaceae)

Common name

Eng.: Emblic, Indian goose berry
Hindi: Amla
Beng.: Amloki.

Distribution

A moderate sized deciduous tree native to S. E. Asia and is now distributed throughout India.

Parts used: Fruits.

Pharmacological activities

Raw fruits are used medicinally as diuretic and laxative and dried from in diarrhea and dysentery. Phyllemblin obtained from fruit pulp has been found to have mild depressant action on central nervous system. Fruit is considered to be good liver tonic. Aqueous extract of the fruits of the plant can prevent toxic effect of lead nitrate $\{Pb(No_3)_2\}$ and alluminium sulphate $\{Al_2(SO_4)_3\}_3$, $18H_2O$ on liver parenchyma cells.

Chemical constituents

Phyllemblin obtained from fruit pulp. Seed fat contained linoleic acid (64.8%) and closely resembled linseed oil. Ellagic acid and lupeol was isolated from roots. Trigalloyl-glucose, terchebin, corilagin, ellagic acid from fruits.

18. *Euphorbia nematocypha* Forst. (Euphorbiaceae)

Parts used: Whole plant.

Pharmacological activities

Significant hepatoprotective activity has been found due to presence of brevifolin, hyperin, ellagic acid and 3, 3′-di-O-methyl-ellagic acid in the plant. The hepatoprotective activity of these compounds has been ascribed to their antioxidative effects.

Chemical constituents

Brevifolin, hyperin, ellagic acid and 3,3-di-O-methyliellagic acid have been isolated from the plant.

19. *Fructus schisandrae* (Turcz.) Baill. or (Magnoliaceae)

Common name

Syn.: *Schisandra sphenanthera* Rehd. et Wils.
Eng.: Chinese magnoliavine fruit.

Distribution

Northern Wuweizi of specie *Schisandra chinensis* (Turcz.) Baill. is mainly produced in Jilin, Liaoning, Heilongjiang and Hebei provinces of China.

Southern Wuweizi of specie *Schisandra sphenanthera* Rehd. et Wils. is mainly produced in Hubei, Henan, Shaanxi, Shanxi, Gansu and Sichuang provinces of China.

Parts used: Kernel.

Pharmacological activities

Significant hepatoprotective activity has been reported from the Kernel of the plant.

Chemical constituents

Lignans belonging to dibenzocylo octeno group isolated from the alcoholic extract of the kernel of the plant. Northern Wuweizi (*Schisandra chinensis*) fruit contains 0.89% volatile oils. They include sesquicarene, B2-bisabolene 0-β-Chamingrene and α-ylangene. It also contains approximately 5% lignans. It contains 9.11% organic acids, which include citric acid, malic acid, tartaric acid, succinic acid and vitamin C. The seeds contain 33% fatty oil.

Lignans include schisandrin, and its derivatives alpha, beta, gamma and Theta, schisandrin, pseudo-schisandrin, deoxy-schisandrin, neoschisandrin, schisandrol and others.

In addition to the above isolated lignan ingredients, the following lignan ingredients are also isolated. They include gomisins A, B, C, D, E, F, G, H, J, N, O, (-)-gomisin K1, (+)-gomisin K2, (+)-gomisin K3, (-)-gomisin L1, (-)-gomisin L2, (+)-gomisin M1, (+)-gomisin M2, epigomisin O, angeloylgomisin Q, angeloylgomisin P and tigloylgomisin P. The last two compounds are optical isomers.

Other compounds include citral, chlorophyll, sterols, vitamins C, E, resins, tannin and little amount of sugars.

20. ***Garcinia kola*** **Heckel. (Clusiaceae)**

Distribution

Tropical, Asia, Africa.

Pharmacological activities

Kolaviron isolated from the plant was found to produce significant hepatoprotective effect against CCI_4 induced liver injury. Later studies showed that Kolaviron, a mixture of bioflavonoids and kolaflavanone isolated from the plant against thioacetamide induced liver toxicity in rats which was measured by thiopental sleeping time indicating significant hepatoprotective activity. Besides, they were also found to alter serum microsomal enzyme levels.

Chemical constituents

Kolaviron, a mixture of bioflavonoids and Kalaflavanone has been isolated from the plant. B-sitosterol isolated from seeds; lipid from seed consisted of unsaturated and saturated acids; stearic, palmitic, myristic and oleic acids identified.

21. ***Garcinia mangostana*** **L. (Clusiaceae)**

Common names
Eng.: Mangosteen
Hindi: Mangustan
Beng.: Mangustan

Distribution

Tree native to Malaysia. Mangosteen fruits are chiefly imported into India from Singapore and the strait settlements, through to some extent it is cultivated in Madres.

Parts used: Fruits.

Pharmacological activities

Fruit is used in chronic diarrhea and dysentery. It also possesses hepatoprotective activity.

Chemical constituents

Three new xanthones gartanin, deoxygartanin and normangostin was isolated from isolated from fruits. Mangostin was isolated and its structure was confirmed.

22. ***Glycyrrhiza glabra*** **Linn. (Fabaceae)**

Common names
Eng.: Liquorice
Beng.: Jashtimadhu

Distribution

Persian gulf, Asia Minor, Turkestan, Siberia, China, France, Germany, Italy.

Parts used: Roots.

Pharmacological activities

Glycyrrhizin analogues (Glycyrrhizin glycyrrhetinic acid) have been isolated from the plants which possess hepatoprotective activity.

Chemical constituents

See Antifertility Chapter.

23. ***Gymnosporia montana*** **Roth. (Celastraceae)**

Syn. **G. spinosa**

Common names
Hindi: Vingar
Beng.: Vaichigachha
Sans.: Vikankata

Distribution

Tropical, Subtropical and Africa.

Parts used: leaves.

Pharmacological activities

Leaf extract of the plant showed hepatoprotective activity. Primary screening was carried out in mice by noting the effects on CCI_4 induced prolongation of pentobarbitone sleep. Extract found effective in the primary test, was further evaluated in rats by noting their effect on CCI_4 induced alterations in morphological, biochemical and histopathological parameter. Methanolic extract of defatted leaf (ME) of the plant produced significant reversal of majority of the altered biochemical parameters stud led and histopathological studies confirmed the presence of significant hepatoprotective activity in the extract.

Chemical Constituents

Tingenone, 3-O-acetyloleanolic acid, hexacosane, betulin, B-amyrin, hexacosanol and B-sitosterol isolated from stem bark root bark land leaves.

24. ***Hypoestes triflora*** **L. (Acanthaceae)**

Part used: leaves.

Pharmacological activities

Treatment with aqueous extract of the leaves prevented CCI_4, induced prolongation of duration of barbiturate sleep, indicating presence of hepatoprotective activity in it. The active principle responsible for the hepatoprotective activity is reposted to be benzoic acid. It presented CCI_4induced elevation in transaminase activity in mice.

Chemical constituents

Benzoic acid.

25. ***Indigofera tinctoria*** **Baker. (Fabaceae)**

Common names
Hindi: Neel
Beng.: Neel
Tam.: Nils

Distribution

Available in W. Bengal, Bihar, Tamilnadu and Karnataka.

Parts used: Aerial parts.

Pharmacological activities

A detailed investigation of the alcoholic extract of the aerial part of the plant has been done. The extract was found to afford significant protection to mice, rats and rabbits against CCI_4 induced hepatic injury. Parameters measured were pentobarbitone sleeping time, prothrombin time and bromsulphalein excretion. It was also found to increase liver weight and bile flow in rats indicating microsomal enzyme induction.

Chemical constituents

Analysis of the leaves gave the following values (dry basis); Nitrogen (N_0, 5,11; Phosphoric acid (P_2O_5), 0.18; Potash (K_2O), 1.67; and lime (CaO) 5.35%.

26. ***Lawsonia alba*** **Lam. (Lythraceae)**

Common name
Eng.: Henna
Hindi: Mehandi
Beng.: Mehandi

Distribution

A much braced shrub or small tree native to Arabia and Persia and is now cultivated in Haryana, Gujarat and to a small extent in M.P. and Rajasthan.

Parts used: Barks.

Pharmacological activities

Hepatoprotective activity has been reported in ethanol water extract of the bark of the plant against CCI_4 induced liver toxicity. The extract was effective in preventing CCI_4 induced prolongation of hexobarbitone induced sleep and alteration in biochemical parameters and BSP clearance.

Chemical constituents

Laxanthones I and II isolated and characterized as 1,3-dihydroxy-6,7-dimethoxyxanthone and 1-hydroxy-3,6-diacetoxy-7-methoxyxanthone respectively. Laxanthone III isolated and characterized as 1-hydroxy-3,7-dimethoxy-6-acetoxyxanthone.

27. *Liquidambar formosana* Hance. (Hamamelidacea)

Common names
Eng.: Fragrant maple

Distribution

A tree native to China now reported to have been introduced in Lalbagh gardens, Bangalore.

Parts used: Fruits.

Pharmacological activities

Hepatoprotective activity has been reported in Betulonic acid, an isolated from fruit of the plant.

Chemical constituents

Betulonic acid has been isolated from fruit of the plant.

28. *Luffa echinata* Roxb. (Cucurbitaceae)

Common names
Punjab: Bakru
U.P.: Takhoi
Kashmir: Tata bateri
Beng.: Bindal

Distribution

Hills of Kashmir, U.P. and Kumooni.

Parts used: Fruits.

Pharmacological activities

Alcoholic and ether extracts of the plant were found to protect rats against CCI_4 induced injury. From the extract of the plant, echiantin, cucurbitacin B, a spooning glycoside of gypsogenin, a flavonoid-chrysoeriol, flavones epigamic and fuelling have been isolated by different workers, but which one of them is responsible for hepatoprotective activity remains to the determined.

Chemical constituents

Echinatin, cucurbitacin B, a saponifying glycoside of gypsogenin, a flavonoid-chrysoeriol, flavones apigenin and futeolin have been isolated from alcoholic and ether extracts of the plant. Cucurbitacin B (3,5), m.p. 176^0 cucurbitacin E (0.05%) mp. 234^0 and a bitter saponin composed of a triterpene acid, glucose arabinose and rhamnose, from fruits; cucurbitacin B, B-sitosterol two unidentified triterpene alcohols echinatol a;

mp. 144^0 and echinatol B, mp. 167^0 from seeds; constitution of cucurbitacin B established; structures of cucurbitacin A and C proposed.

29. *Nymphaea stellata* Willd. (Nymphaeaceae)

Common names
Eng.: Indian blue water lily
Hindi: Nilpadma
Beng.: Nilshapla

Distribution

Native of South East Asia, aquatic herb found in ponds and ditches.

Parts used: Whole plant.

Pharmacological activities

Petroleum ether extract of the plant provides significant protection to rats against CCI_4induced functional, histopathological and morphological changes. The extract was found to promote parenchymal tissue regeneration.

Chemical constituents

Analysis of the dried tubers gave the following values; moisture, 4.20; fat, 0.45; proteins, 14.56; carbohydrate's 67.49; fibre, 5.45 and ash, 7.85%.

30. *Ocimum sanctum* Linn. (Lamiaceae)

Common names
Eng.: Sacred basil
Hindi: Tulsi
Beng.: Tulsi

Distribution

A well known sacred plant of India grown in gardens, temples, houses all over India.

Parts used: Leaves.

Pharmacological activities

Extracts protected rats against CCI_4 induced liver injury.

Chemical constituents

It is reported to contain ursolic acid, apigenin, luteolin, apigenin-7-O-glucuronide, luteolin-7-O-glucuronide orientin and molludistin. However, no study has been made to evaluate hepatoprotective activity of these compounds. Eugenol (70.5), its methyl ether, nerol (6.4), caryophyllene (7.5), terpinen-4-ol (0.4), decyoladehyde (0.2), ψ-selinene (0.4), β-pinene (0.4), camphene (2.0) and α-pinene (3.5%) identified in essential oil by GC.

31. *Paederia foetida* Linn. (Rubiaceae)

Common names
Hindi: Ghandhali, Somraji
Beng .: Gandhal
Sans.: Prasarani

Distribution

Found in Central and Eastern Himalayas, extending to Calcutta.

Parts used: Leaves.

Pharmacological activities

Leaf extract of the plant responded hepatoprotective activity. Primary screening was carried out in mice by noting the effects on CCI_4induced prolongation of pentobarbitone sleep. Extracts effects were evaluated in rats by noting their effect on CCI_4 induced alterations in morphological, biochemical and histopathological parameters. Methanolic extract of the plant also showed reversal of some of the altered biochemical parameters. However, the extract did not prevent CCI_4induced histopathological changes to significant extent.

Chemical constituents

Asperuloside and four new iridoid glucosides-paederoside, m. p. 122^0, paederosidic acid, scandoside, m. p. 139^0 and deacetyl asperuloside were isolated and characterized. Hentriacontane, hentriacontanol, methyl mercaptan, ceryl alcohol, palmitic acid, sitosterol stigmasterol, campesterol, ursolic acid and iridoid glycosides-asperuloside, paederoside and scandoside isolated from leaves and stems.

32. *Peumus boldus* Molina. (Monimiaceae)

Parts used: Whole plant.

Pharmacological activities

It has been reported the presence of hepatoprotective activity in dried hydro alcoholic extract of the plant against tert-Butylhydroperoxide induced cytotoxicity in isolated rat hepatocytes. It also protected mice against CCI_4 toxication (*in vitro*). Boldine the main alkaloid of the plant has been implicated for the hepatoprotective activity.

Chemical constituents

An alkaloid boldine has been isolated.

33. *Phyllanthus amarus* Linn. (Euphorbiaceae)

Distribution

South India.

Parts used: Whole plant.

Pharmacological activities

In an interesting study, it has been reported hepatitis B virus inactivating activity in the alcoholic extract of the plant in an *in vitro* test system (unpublished work).

34. *Phyllanthus fraternus* Webster. (Euphorbiaceae)

Syn. P. niruri Sensu Hook. F. (non L.)

Common names
Sans.: Bhumyamalaki
Hindi: Jar-amla
Beng.: Bhui-amla

Distribution

Throughout the hotter parts of India from Punjab to Assam and Southwards to Travancore, ascending to the hills to 2000 ft.

Parts used: Whole plant.

Pharmacological activities

The plant is used as remedy for jaundice.

Chemical constituents

The plant contains the bitter substances phyllanthin and hypophyllanthin. The phyllanthin was characterized as (+) 3,3′, 4,4′, 9,9′ hexamethoxy 8,8′ butyrolignan. Three new lignans niranthin; nirtetralin and phyltetralin isolated from leaves; detection of estradiol in bark and roots by TLC estradiol contents 155-350 ug/100g in plant samples; Kaempferol-4′-rhamnopyranoside and eriodictyol-7-rhamnopyranoside isolated from roots; lup-20-en-3Beta-ol and its acetate isolated from roots.

35. *Picrorhiza kurroa* Royle ex Benth. (Scrophulariaceae)

Common names
Hindi: Kuru
Beng.: Kutki
Sans.: Katuka

Distribution

The Himalayas from Kashmir to Sikkim.

Parts used: Roots.

Pharmacological activities

CDRI group through an elaborate study employing a number of biochemical parameters, has evaluated hepatoprotective activity in Picroliv, a standardized fraction of the root of the plant containing about 60% of a mixture of picroside I and Kutkoside in the ration of 1:1:5. The test drug significantly reversed alterations in number of biochemical parameters caused by CCI_4 administration. Noteworthy among these parameters quantity in normal rats. In the drug treated group it was lowered significantly and could not be measured. The overall hepatoprotective activity was evaluated by measuring 'PAV' value, *i.e.*, percentage proportion of abnormal value, which was found to be significantly less in Picroliv treated group in comparison with CCI_4 control rats.

Chemical constituents

Picroliv which is a mixture of picroside I and Kutkoside in the ratio 1:1:5 has been obtained from the root of the plant. A new iridoid glucoside-picroside II isolated and characterized as 6-vanilloylcatalpol. Picroside 111 was isolated and characterized as 6′-(4-hydroxy-3-methoxy cinnamoyl) catalpol.

36. *Piper longum* Linn. (Piperaceae)

Common names
Eng.: Long pepper
Hindi: Pipal
Beng.: Piplamur

Distribution

Cultivated in West Bengal, Karnataka and Tamilnadu, also in plains of Bengal.

Pharmacological activities

Milk extract (decoction with milk) of the plant has been tested for hepatoprotective activity against CCI_4 induced hepatic injury. The extract though ineffective in preventing acute changes, promoted regeneration process by restricting fibrosis.

Chemical constituents

See Antifertility Chapter.

37. *Quercus infectoria* Olivier. (Fagaceae)

Common names
Hindi: Maza, Muphal
Beng.: Majuphal
Eng.: Gak oak

Distribution

A shrub or small tree indigenous to Greece, Syria and Iran.

Parts used: Fruits.

Pharmacological activities

Some co-workers have screened fruit of the plant for hepatoprotective activity against CCI_4 induced liver damage. Ethanolic extract to the fruit of the plant was found to afford significant protection.

Chemical constituents

Methyl betulate, methyl oleanolate sitosterol, syringic, gallic and ellagic acids were isolated from the plants.

38. *Ricinus communis* Linn. (Euphorbiaceae)

Common names
Eng.: Castor seed, castor, castorbean, castor oil plant
Hindi: Arandi
Beng.: Rerhi

Distribution

A small tree cultivated chiefly in Andhra Pradesh, Maharashtra, Karnataka, Orissa, available in wild condition in W. Bengal.

Parts used: Roots.

Pharmacological activities

Crude extracts and its butanol soluble fraction of the plant have been reported to afford significant protection to rat's against Gal-N induced hepatic damage.

Chemical constituents

Palmitic (1.2) Stearic (0.7), arachidic (0.3). hexadecenoic (0.2), oleic (3.2), linoleic (3,4), linolenic (0.2) ricinoleic (89.4%) and dihydroxystearic acid as Me-esters in castor oil was detected. Lupeol and 30-norlupan-3B-ol-20-one from coat of castor bean was reported.

39. *Salsola collina* Linn. (Chenopodiaceae)

Distribution

North-East Himalaya.

Parts used: Shoot.

Pharmacological activities

Ethanolic extract of shoot (above ground part) was found to provide good protection against CCI_4 induced changes in metabolic functions and liver cytoarchitecture.

40. *Sambucus formosana* Rehder. (Caprifoliaceae)

Common names
Eng.: Elder

Distribution
Eastern Himalaya.

Parts used: Whole plant.

Pharmacological activities

Significant hepatoprotective activity has been observed against CCI_4 induced liver injury. The hepatoprotective activity was due to sambuculin A isolated from the plant. From the same plant Beta-amyrin and oleanolic acid were also isolated. A mixture of α-amyrin and β-amyrin palmitate exhibited strong hepatoprotective effect CCI_4 induced liver injury.

Chemical constituents

Sambuculin A, Beta-amyrin and oleanolic acid have been isolated.

41. *Silybum marianum* L. (Asteraceae)

Common names
Eng.: Milk thistle.

Distribution

Widely distributed in S. America.

Parts used: Seeds.

Pharmacological activities

Reported to have hepatoprotective activity.

Chemical constituents

Optically active dehydrodiconifenyl alcohol isolated from seeds; twelve polyacetylenes and one polyene detected in roots; structure of seven of these determined whereas other substances could only be partially characterized; and isomer of substances could only be partially characterized; and isomer of silymarin silychristin-isolated from fruits and characterized; myristic, palmitic, stearic and oleic acids, taxifolin, silybin and silydianin, silibonol isolated from seeds; a new flavonolignan-2,3-dehydrosilymarin and 2,3-dehydrosilychristin isolated from seeds; silymonin and silandrin isolated from fruits; tannins deremined in flowers, leaves and stem.

42. ***Solanum incanum*** **Kuntze. (Solanaceae)**

Syn. **S. Coagulans** Forsk.

Common names
Hindi: Congulus forsk,
Punjab: Barimalehari

Pharmacological activities

Hepatoprotective activity was assessed by CCI_4 induced pentobarbitone (PBN) sleep prolongation and elevation of transaminases activity. Carpesteral an isolate, completely prevented PBN sleeping besides lowering transaminase activity to almost normal level.

Chemical constituents

Carpestrol has been isolated from the plant, besides Solasodine, solamargine and solasonine and ursolic acid. Diosgenin and yamogenin were isolated in 1.2% yield from fruit.

43. ***Tinospora cordifolia*** **Willd. (Menispermaceae)**

Common names
Hindi: Gulancha, Giloe
Beng.: Gadancha
Sans.: Amrita

Distribution

A climbing shrub distribution throughout tropical India and Andamans.

Parts used: Dry stems with bark.

Pharmacological activities

Biochemical, morphological and histopathological parameters have been evaluated from the decoction of the plant and hepatoprotective activity against CCl_4 induced hepatic injury has been tested.

Chemical constituents

Two bitter substances, substance A ($C_{22}H_{34}O_{10}5H_2O$) m.p. 226^0-228^0 substance B (m.w. 495 melts 186-188^0). Giloin, $C_{32}H_{32}O_{10}H_{20}$, a glycoside, mp. 226-228^0, Gilenin $C_{17}H_{18}O_5$, a non-glycoside bitter mo. 210-212^0 and Gilo sterol, $C_{28}H_{48}O$, mp. 192-93^0 were isolated.

44. ***Tephrosia purpurea*** **Pers. (Fabaceae)**

Common names
Sans.: Sharpunkha
Hindi: Dhamasia
Beng.: Ban-Nil-gachh

Distribution

It is a sub-erect herbaceous perennial found all over India, ascending to the Himalayas upto an altitude of 6000 ft.

Parts used: Whole plant.

Pharmacological activities

Hepatoprotective activity has been observed.

Chemical constituents

The roots contain tephrosin, deguelin, isotephrosin, rotenone etc. The leaves contain about 2% a glycoside, osyritin, caffeic acid isolated from dormant seeds; rutin, B-sitosterol and lupeol isolated from leaves; delphinidin chlordin and cyanidin chloride isolated from flowers.

45. *Thujopsis dolabrata* Sieb. & Zucc. (Cupressaceae)

Distribution

Japan.

Parts used: Whole plant.

Pharmacological activities

Hepatoprotective activity in desoxy-podophyllotoxin isolated from the plant has been observed.

Chemical constituents

Desoxy-prodophyllotoxin has been isolated from the plant.

46. *Vitex negundo* L. (Verbenaceae)

Common names
Hindi: Sambhalu
Beng.: Nishinda
Sans.: Nirgundi

Distribution

A shrub or smell tree found in W. Bengal.

Parts used: Seeds.

Pharmacological activities

Significant hepatoprotective activity in the alcoholic extract of the plants seeds has been observed. The effect was found to be significantly reverse CCI_4 induced changes in morphological, biochemical and functional parameters.

Chemical constituents

n-tritriacontane, n-hentriacontane, n-pentatriacontane, n-nonacosane, Beta-sitosterol, p-hydroxybenzoic acid and 5-oxyisophthalic acid from seeds; 3,4-dihydroxybenzoic acid also isolated; vanillic and p-hydroxyanidins isolated from stem bark and their structures determined as 6,8-di-O-methyleucodelphinidin and 3'4'-di-O-methyleucocyanidin-7-O-rhamnoglucoside.

47. *Wedelia chinensis* Merrill. (Asteraceae)

Common names
Hindi: Bhangra
Beng.: Bhimra
Sans.: Bhringaraja

Distribution

A herb found in U.P., Assam, Arunachal Pradesh and W. Bengal.

Parts used: Leaves.

Pharmacological activities

Hepatoprotective activity has been observed in the methanol extract of the plant. Extract tested against CCI_4 and Gal-N induced cytotoxicity in primary cultured hepatocytes.

Chemical constituents

Isolation of wedel lactone.

48. *Withania frutiscens* Linn. (Solanaceae)

Parts used: Leaves.

Pharmacological activities

Ethanolic extract of the leaves of plant has been reported to prevent CCI_4 induced alteration in pentobarbitone sleep, biochemical parameters studied and derangement of liver cytoarchitecture has been observed.

49. *Withania somnifera* Dunal. (Solanaceae)

Common names
Eng.: Ashwagandha
Hindi: Asgandh
Beng.: Ashwagandha

Distribution

This is an erect shrub found throughout the drier parts of India, Baluchistan and Ceylon.

Parts used: Roots and tuber roots.

Pharmacological activities

Alcoholic extract of the leaves of the plant was found to significantly inhibit CCI_4 induced alterations in transaminase activity and pentobarbitone sleeping time indication presences of hepatoprotective activity. This was confirmed through histopathological studies.

Chemical constituents

See Antifertility Chapter.

50. *Withania coagulans* Dunal. (Solanaceae)

Common names
Eng.: Vegatable rennel, Indian Cheese-maker
Hindi: Akri, Punir
Tel.: Panneru-gadda

Distribution

A small shrub found in Punjab.

Parts used: Fruits.

Pharmacological activities

Fruits are used in liver complains. A withanolide, 3-B-hydroxy-3,3-dihydro withanolide f, isolated from the fruit of the plant also showed significant hepatoprotective activity against CCI_4 induced liver injury. The activity was assessed by measuring pentobarbitone sleeping time, serum transaminase activity and through histopathological studies.

Chemical constituents

Different compounds (with melting points 125^0, 90^0 81^0 and 128^0) was isolated from the fruits. A new steroid-5,27-dihydroxy-6a,7a-epoxy-1-oxo-5a-with a-2,24-dienolide along with withaniol and withaferin A was isolated. A withanolide, 3-b-hydroxy-3,3-dihydro withanolide F was isolated from fruits of the plant.

6

MEDICINAL PLANTS FOR ANTIDIABETIC AGENT

INTRODUCTION

Diabetes

Diabetes is a chronic disorder of metabolism of carbohydrates, proteins and fats due to absolute or relative deficiency of insulin secretion and with varying degree of insulin resistance.

Historical Perspective

Diabetes is a disorder known from ancient civilization. Diabetes has been mentioned in Ayurveda by Sushruta as a state in which urine is sweet. He described diabetes as a condition, 'Madhumeha' in which person passes urine, which resembles honey. Sushruta had advised starvation and avoidance of sweet food substances for a person suffering from madhumeha.

The word 'diabetes' was coined by the Egyptian Papyrus Ebers nearly 3500 years ago, who described the symptoms of diabetes are excessive thirst, weight loss, excessive hunger and presence of glucose in the urine.

In 17th century, Willis observed that the urine of diabetics was "wonderfully sweet as if imbused with honey or sugar". The presence of sugar in urine of diabetics was demonstrated by Dobson in 1755.

In 1869 *Paul Langerhans* observed that the pancreas consists of two types of cells- the acinar cells, which secrete digestive enzymes, and cells clustered like islands, or islets, which had an unknown function. Direct evidence for this function came in 1889, when *Oskar Minkowski* and *Joseph Von Mering* showed that pancreatic tomized dogs exhibit a syndrome similar to *diabetes mellitus* in human beings. During the early part of this century, several investigations were close to the discovery of the pancreatic substance that reduces blood glucose. In 1921, *Banting* and *Best*, successfully obtained a pancreatic extract that was effective in decreasing the concentration of blood glucose in diabetic dogs.

There after in 1922, a pancreatic extract was administered to a 14-year-old patient exhibiting a serum glucose level of 500 mg/dl, who survived 15 years on insulin obtained from animal source. This success led many to conclude, erroneously, that *diabetes* is a disease of elevated blood sugar, easily corrected by exogenous insulin.

The chemical structure of insulin was elucidated by *Sanger* in 1960, and this led to the complete synthesis of the protein in 1963 and the insulin molecule was fully synthesized in the laboratory by 1966. In 1972, its 3D structure was determined by *Hodgkin* and *Coworkers* in 1972.

Types of *Diabetes*

Diabetes is a condition which chiefly encompassing a large urine volume described by Greek and Roman physicians and classified into 2 types.

1. *Diabetes insipidus*

(*Diabetes*=overflow; *insipidus*=taste less) Two types of problems can cause diabetes insipidus.

(a) Neurogenic (central) *diabetes insipidus* results from hyposecretion of ADH usually caused by brain tumor, head trauma or brain surgery that damages posterior pituitary gland and the hypothalamic para ventricular and supraoptic nuclei.
(b) In Nephrogenic *diabetes insipidus* the kidneys do not respond to ADH. The ADH receptor may be non functional or do not respond to ADH. ADH receptors may be non functional or the kidneys may be damaged.

Symptoms of diabetes insipidus include excretion of large volumes of urine, with resulting dehydration and thirst.

2. *Diabetes mellitus*

It is a chronic metabolic disorder, resulting from insulin deficiency, characterized by hyperglycaemia, altered metabolism of carbohydrates, proteins and lipids, and an increased risk of vascular complications. The insulin deficiency may be absolute or relative and the metabolic abnormalities lead to classic symptoms of polyuria (frequent urination), polydipsia (excessive thirst), polyphagia (excessive hunger) and fatigue. Long-term complications of *Diabetes mellitus* include gangrene, proliferative retinopathy and uremia. The disordered micro circulation (micro angiopathy) underlies the majority of these complications.

Most of the Pathological features of diabetes mellitus can be attributed to one of the following major effects of insulin lack:

1. Decreased utilization of glucose by the body cells, with resultant increase in blood glucose concentration to 300 to 1200 mg/dl;
2. Markedly increase mobilization of fats from the fat storage areas, causing abnormal fat metabolism as well as deposition of cholesterol inarterial walls causing atherosclerosis; and
3. Depletion of proteins in tissues of the body.

Types of Diabetes Mellitus

Two types of *diabetes mellitus* are

Type I Diabetes mellitus

Insulin dependent *diabetes mellitus* (IDDM), juvenile onset *diabetes mellitus*; there is cell destruction in pancreatic islets; majority of cases are auto immune (type 1A) antibodies that destroy b cells are detectable in blood, but some are idiopathic (type 1B) – no b cell anti body is found. In all type 1 cases circulating insulin levels are low or very low, and patient are more prone to ketosis. It previously was known as juvenile onset diabetes because it most commonly develops in younger than age 20, although persist throughout life.

The main symptoms include high blood sugar, excessive thirst, frequent urination, increased appetite, weight loss, poor wound healing, blurred vision etc. Since insulin is not present to aid the entry of glucose into body cells, most cell use fatty acids to produce ATP. Organic acids called ketones are by products of fatty acid catabolism. As they accumulate, they cause a form of acidosis called keto acidosis, which lowers the pH of the blood and can result in death. The catabolism of stored triglyceride and proteins also causes weight loss. As lipids are transported by the blood from storage depot to cells, lipid particles are deposited on the wall of blood vessels. The deposition leads to atherosclerosis and multitude of cardiovascular problems including cerebrovascular in sufficiency, ischemic heart disease, peripheral vascular disease and gangrene. One of major complication of diabetes is loss of vision due to cataract or damage to blood vessels of retina.

Type II Diabetes mellitus

Non insulin-depenent *diabetes mellitus* (NIDDM), type 2 diabetes is, characterized by insulin resistance and impaired insulin secretion. It usually develops after age 40 and is not associated with total loss of the ability to secret insulin. This is associated with variety of symptoms like polyphagia, polyuria, polydypsia. Other symptoms include weight loss, poor wound healing, dry mouth, itchy skin, blurred vision, impotence, recurrent infection, hypertension, atherosclerosis etc. Major chronic complication include accelerated macro vascular disease, retinopathy, renal disease, and neuropathy.

Causes of NIDDM

Abnormality in gluco-receptor of β cells so that they respond at higher glucose concentration.

Reduced sensitivity of peripheral tissues to insulin: reduction in number of insulin receptors, 'down regulation' of insulin recepters. Many hypertensives are hyper insulinemic but normoglycemic; exhibit insulin resistance. Hyper insulinemic *per se* has been implicated in causing angiopathy.

Excess of hyperglycemic hormones (glucagons etc) /obesity cause relative insulin deficiency-the β cells lag behind.

Therapy of Diabetes mellitus

Insulin

Insulin is the hormone secreted by the β-cells of islets of Langerhans of pancreas. It derives its name from the latin word 'insula' which means an island. Insulin was First isolated in 1921 by Banting and Best, and used in treatment of diabetes mellitus in 1922. It was completely synthesized in 1996. Insulin was the first protein proved to have hormonal action, first protein crystallized, the first protein sequence, the first protein synthesized by chemical techniques, the first protein shown to be synthesized as a large precursor molecule and the first protein prepared for commercial use by recombinant DNA technology. The chemical structure of Insulin was elucidated by *Sanger* in 1960 and the insulin molecule was fully synthesized in the laboratory by 1966.

Insulin is a polypeptide with a molecular weight of about 6,000. It is composed of an A-chain (acidic) made up of 21 amino acids, and B-chain (basic) of 30 amino acids, linked by two disulphide (-S-S-) bridges. Chain A and B, linked by 2 intrachain disulphide bridges that connect A7 to B7 and A20 to B19. A third intrachain disulphide bridge connects residue 6 and 11 of A chain. Molecular weight of human insulin is 5734.

Biosynthesis of insulin

Rubenstein and Steiner in 1971 demonstrated that insulin is released in β cell of pancreases through enzymatic conversion of pro insulin. It is a single chain, high molecular weight-peptide and is devoid of any biological activity. A thiol activated protease enzyme catalyzes this conversion. Chance *et al.*, in 1976, going a step ahead, proposed the existence of precursor of even proinsulin, which named as prepro insulin. The protease enzyme catalyses the cleavage of proinsulin molecule resulting into formation of insulin, four basic amino acids (arginine 31, arginine 32, lysine 64 and arginine 65) and one molecule of C-peptide (the connector containing amino acids from 33 to 63).

Preproinsulin (109 amino acids) is synthesized in the ribosomes of endoplasmic reticulum of pancreatic β-cells. At the same site, conversion of preproinsulin to proinsulin occurs. The latter is transferred through vesicals to the Golgi apparatus and stored in immature granules. The conversion of proinsulin to insulin is slow and gradual process that continues along with maturation of storage granules. As most of the insulin is get converted to insulin (about 4-5 hours), the conversion products and starting material get crystallized with zinc. There after, it may be released by exocytosis or may be stored further. The release is signaled by neuronal and or hormonal factors.

Glucose transporters

Insulin is secreted, in response to elevated serum glucose level, by β-cells of the pancreatic islets of Langerhans. The role of insulin is to stimulate GLUT-4 glucose transporter. GLUT-4 is the most important of the glucose transporter molecule and by insertion into the muscle and adipose cell membranes serves to facilitate glucose delivery into these cells. Among the glucose transporter, GLUT-1 transporter govern the basal glucose uptake into all cell, but especially brain and red cells, where as GLUT-2 and GLUT-3 regulate insulin release and insulin uptake mostly into neurons and kidney cells respectively. GLUT-4 is the principle insulin sensitive glucose transporter. GLUT-2 and GLUT-3 are not governed by insulin, and GLUT-3 has a Michaelis constant value (<1 mM) lower than for GLUT-4 (5 mM), so that is available at low serum insulin level.

Regulation of insulin secretion

Under basal condition H"1U insulin is secreted per hour by human pancreas. Secretion of insulin from β-cells is regulated by chemical, hormonal, and neural mechanism.

Chemical

The β-cells have a glucose sensing mechanism depend on entry of glucose into β-cells. Activation of gluco receptor indirectly causes partial depolarization of the β-cells and increases intracellular Ca^{2+} availability (due to increased influx, decreased efflux and release from intracellular stores) Exocytotic release of insulin.

Hormonal

A number of hormones, *e.g.*, growth hormone, corticosteroids, thyroxine modify insulin release in response to glucose.

Neural: The islets are richly supplied by sympathetic and vagal nerves. Adrenergic α2 receptor activation decreases insulin release and adrenergic β2 stimulation increases insulin release. Cholinergic- muscarinic activation by Ach or vagal stimulation causes insulin secretion.

Insulin receptor

The receptor which has molecular weight approximately 340,000 is a tetramer made up of 2α and 2β glycoprotein subunits. All these are synthesized on a single m RNA and then proteolytically separated and bound to each other by disulphide bonds. The α subunits bind insulin and are extracellular, where as the β subunits span the membrane. The α subunits carry insulin binding site while the β subunits have tyrosine protein kinase activity. When insulin binds to the receptor several events occur.

1. There is conformational change of the receptor.
2. The receptor crosslink and form microaggregate.
3. The receptor is internalized and
4. One or more signal is generated.

In condition in which plasma insulin levels are high, *e.g.,* Obesity or acromegaly, the no of insulin receptors is deceased and target tissue become less sensitive to insulin. This down regulation results from the loss of receptor by internalization, the process where by insulin-receptor complexes enter the cell through endocytosis in clathrin-coated vesicles.

Effects of Insulin

- Cells Increased transport of glucose, amino acids and K^+ into insulin sensitive cell.
- Stimulation of protein synthesis
- Inhibition of protein degradation
- Activation of glycolytic enzymes and glycogen synthase
- Inhibition of phosphorylase and gluco-neogenic enzymes

The best known is the hypoglycemic effect. The net effect of the hormone is storage of carbohydrate, protein, and fat.

In diabetes mellitus there are disturbances in the metabolism of carbohydrate, fat protein electrolytes and water.

(a) *Carbohydrate metabolism*

Insulin deficiency produces two fundamental defects.

- Reduce entry of glucose into cells,
- Increased release of glucose from the liver into circulation.

Both of these raise the blood glucose level (hyperglycemia), and lead to excretion of glucose (glucosuria). In addition in insulin deficiency there is an abnormally high rate of conversion of protein to glucose (gluconeogenesis).

(b) *Fat metabolism*

Insulin deficiency leads to a mobilization of fat from adipose tissue into blood stream (lipaemia). This leads to increase in blood concentration of ketone bodies. These ketone bodies enter the blood and produce a *metabolic ketoacidosis*. In diabetes the blood cholesterol level also rises.

(c) *Protein metabolism*

Insulin deficiency and disorder of glucose utilization impair protein synthesis, and promote protein breakdown especially in the muscle. Increased protein catabolism leads to wasting. This protein depletion together with glucose-rich tissue and body fluids predispose individual to infection.

(d) *Electrolytes*

Due to increased protein break down, glycogenolysis, and tissue hypoxia, intracellular potassium escape into the extracellular fluid and is lost in urine.

(e) *Water*

As an osmotic consequence of glucosuria, there is an extra loss of water from the body (polyuia). This leads to the *dehydration* and *polydypsia*.

(A) *Insulin Therapy*

Insulin is the main stay in the treatment of virtually all types of type 1 diabetes mellitus and many type 2 diabetes mellitus. When necessary, insulin may be administered intravenously or intramuscularly; however, long term treatment relies predominantly on subcutaneous injection of the hormone. Preparation of insulin can be classified according to their duration of action into short, intermediate, and long acting.

- Short duration of action (and rapid onset): Soluble Insulin (neutral insulin). *e.g.*, regular insulin, lispro insulin.
- Intermediate duration of action (and slower onset): *e.g.*, Isophane Insulin, NPH, lente.
- Longer duration of action: *e.g.*, Insulin zinc suspension, crystalline, or Protamine Zinc Insulin (glargine, ultralente).

Complications of insulin therapy

(A) Hypoglycemia: Hypoglycemic reactions are most common complication in insulin therapy. They may result from delay in taking meal, unusual physical exertion, or a dose of insulin that is too large for intermediate needs.

(B) *Immunopathology of insulin therapy*: There are two major immune disorders.

1. *Insulin allergy*: It is intermediate type hypersensitivity, is a rare condition in which local or systemic urticaria results from histamine release from tissue mast cell sensitized by anti-insulin IgE antibodies.
2. *Immune insulin resistance*: Most insulin treated patient develops a low titer of circulating IgG anti-insulin antibodies that neutralize the action of insulin in small extent.

(C) *Lipodystrophy at injection sites*: Atrophy of subcutaneous fatty tissue may occur at the site of injection.

(B) Oral Antidiabetic Agents

The oral antidiabetic agents are classified into several categories.

(a) *Sulphonylureas*

Sulphonylureas lower blood glucose level in non-diabetic and Type-2 diabetics. Almost simultaneously they lower the plasma free fatty acid levels. Sulphonylureas stimulate the β-cell islet tissue to secrete insulin, *i.e.*, they have an islet beta-cytotropic activity. Evidence is provided by the finding that they cause degranulation of β-cells. They also probably directly inhibit hepatic glycogenolysis. These are ineffective in Type-1 diabetes as the pancreas almost completely lost its capacity to synthesize and secrete insulin.

(b) *Biguanides*

Biguanides differ from sulphonylureas: cause little or no hypoglycemia in nondiabetic subjects and do not stimulate pancreatic β-

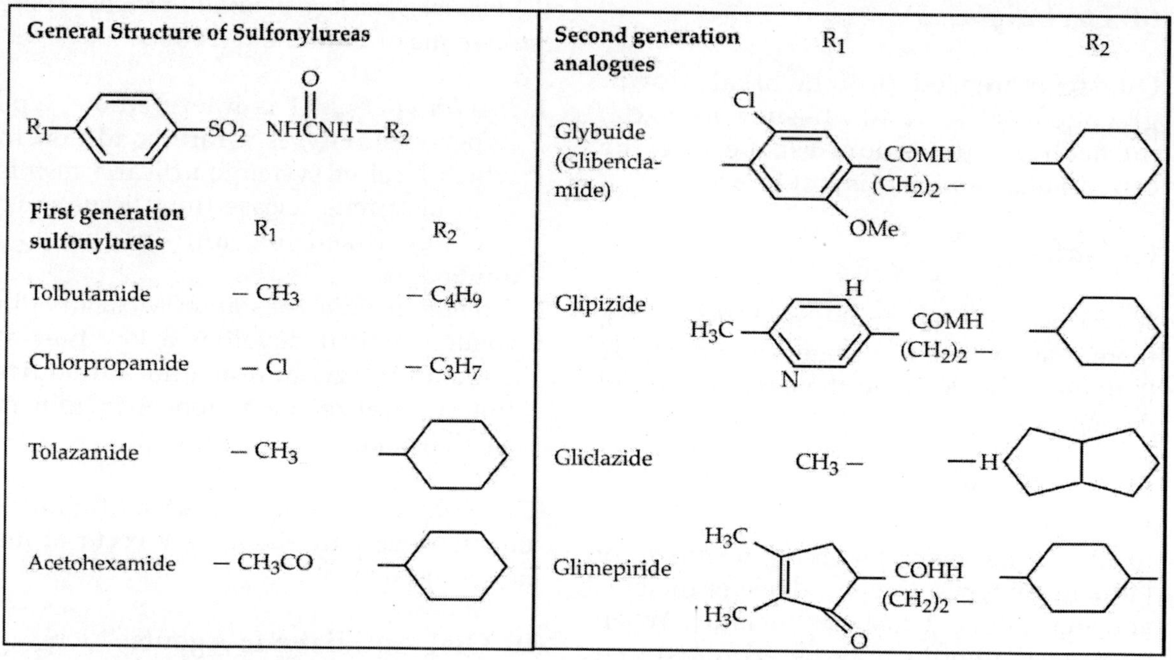

General Structure of Sulfonylureas			Second generation analogues	R1	R2
R_1–C6H4–SO_2 NHC(=O)NH—R_2			Glybuide (Glibenclamide)	Cl, OMe-substituted benzene ring –COMH $(CH_2)_2$ –	–cyclohexyl
First generation sulfonylureas	R_1	R_2			
Tolbutamide	$-CH_3$	$-C_4H_9$	Glipizide	H_3C– pyridine ring (N, H) –COMH $(CH_2)_2$ –	–cyclohexyl
Chlorpropamide	– Cl	$-C_3H_7$			
Tolazamide	$-CH_3$	–cyclohexyl	Gliclazide	CH_3 –	–H bicyclic ring
Acetohexamide	$-CH_3CO$	–cyclohexyl	Glimepiride	H_3C, H_3C cyclopentenone ring (=O) –COHH $(CH_2)_2$ –	–cyclohexyl–

cells. They do not cause insulin release but presence of some insulin is essential for their action. Several mechanisms may be involved in antidiabetic effect of biguanides, they are

- Inhibition of hepatic neoglucogenesis.
- Increase in peripheral glucose utilization.
- Increase in sensitivity of peripheral tissue to insulin
- Delay in glucose absorption from the intestine.
- Reduction in appetite.

General Structure Biguanides

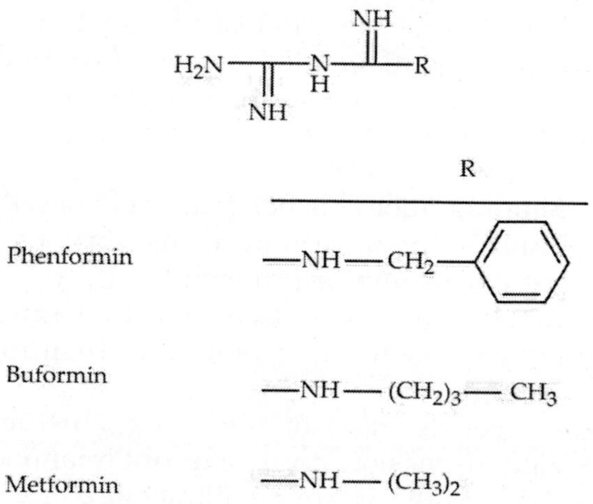

H_2N–C(=NH)–NH–C(=NH)–R

	R
Phenformin	—NH— CH_2—phenyl
Buformin	—NH— $(CH_2)_3$— CH_3
Metformin	—NH— $(CH_3)_2$

(c) *Thiazolidinediones*

These agents are selective agonist of peroxisome proliferator – activated receptor !! *f*—D-*f*—gamma (PPARβ). These agent bind to PPARβ, which in turn, activate insulin responsive genes that regulate carbohydrate and lipid metabolism. These require insulin to be present for their action. They exert effects by lowering insulin resistance in peripheral tissue, lower glucose production by liver, increase glucose transport into muscle and adipose tissue. The thiazolidinediones also can activate genes that regulate free fatty acid metabolism in peripheral tissue.

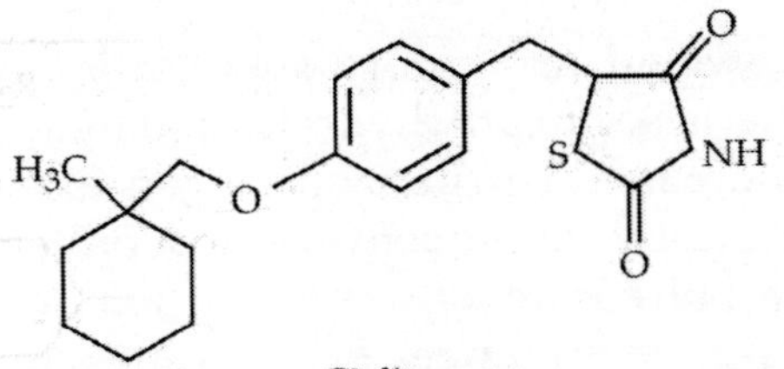

Ciglitazone

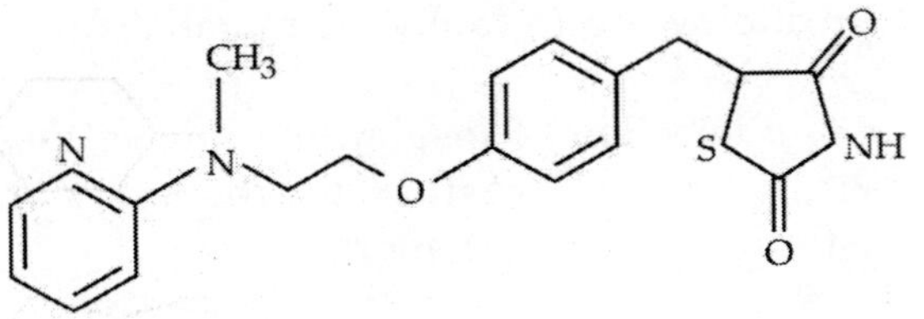

Rosiglitazone

(d) *Alpha-glucosidase inhibitor*

Digestion of complex starches, oligo-saccharides, disaccharides into individual monosaccharides is facilitated by enteric enzymes, including pancreatic α-amylase and α-glucosidases, that are attached to the brush border of intestinal wall. Acarbose and miglitol are competitive inhibitors of α-glucosidases and modulate postprandial digestion and absorption of starch and disaccharides.

A disease of all age-groups practically in all parts of the world: the DIABETES/ DIABETES MELLITUS (Madhumeha), has been well known as a wasting disease due to insulin deficiency in human beings. The pancreas secretes insulin. Carbohydrate metabolism is primarily under the control of insulin. Insulin deficiency occurs in a person due to the functional disorder of the pancreas.

The Pancreas

A mixed gland, the pancreas, a gland of the digestive system, has both exocrine as well as endocrine functions and is therefore called mixed gland. The endocrine part consists of a group of cells forming the Islets of Langerhans which releases hormones related to carbohydrate metabolism and the exocrine part secretes the digestive juice, known as the pancreatic juice, containing proteolytic, lypolytic and amylolytic enzymes.

The pancreas is devoid of distinct connective tissue capsules and is covered by a thin layer of the loose tissue which passes into the gland as septa and subdivides the gland in many lobules. The main pancreatic duct or the duct of Wirsung extends from the left to the right of the organ to open into the duodenum. Within the duodenal well at the ampulla of water, this main pancreatic duct and the common bile duct fuse to form a common duct which opens into the duodenal lumen.

In the pancreas the main pancreatic duct receives many interlobular ducts from each lobule.

The Endocrine Part of The Islets of Langerhans

The normal human adult pancreas contains on an average some 4,000 islets, distributed in scattered manner within the gland, comprising 1 to 3 percent of the total tissue. As each group of cells of the endocrine part is surrounded by the acing of the exocrine part, they look like islands and are hence termed as islets. The distribution of islets is maximum in the tail and minimum in the head of the gland.

Three types of cells are found in the islets. These are called the alpha (a), beta (B) and the delta (d) types. The a cells are fewer in number (about 20%) and they exist peripherally in the islet, while the most numerous B cells (about 75% to 80%) are situated centrally in the form of lumps.

The Syntheses of two hormones insulin and glucagon take place in the B cells and the a cells respectively in the Islet of langerhans. Both hormones play important roles in carbohydrate metabolism. The function of the d cells (about 5% in number) is not clearly known. It is assumed that they may secrete serotonin but some others believe that gastrin is secreted by these cells.

Biosynthesis, Storage, and Catabolism of Insulin

It is already stated that insulin is synthesized by the B cells of Islets of Langerhans but details of the synthesis are unknown. After synthesis, insulin is stored in the B cell of the pancreas and the hormone is released in a controlled and graded manner. The insulin in B cells appears to be associated with granules of the cells. Insulin is released from the B cells into the plasma by the action of glucose and certain other sugars. Growth hormone appears to stimulate insulin secretion. Tolbutamide and other sulfonylureas enhance insulin secretion but C7 sugar, mannoheptulose, may depress its secretion. A number of studies suggest that glucose utilization in the B cells is more or less related to insulin release. Insulin is inactivated in tissues by both enzymic and nonenzymic reductive cleavage of disulfide bonds and by proteolysis. The glutathione reductase enzyme system may be particularly involved in the reductive inactivation.

The Role of Insulin in Carbohydrate Metabolism

By depancreatizing an animal or in diabetes that occurs in human beings, the insulin insufficiency manifests itself in the following manner:

1. Hyperglycemia and glycosuria
2. Depletion of glycogen stores in the liver tissues
3. Lowered respiratory quotient indicating the lack of carbohydrate oxidation.
4. Increase in urinary nitrogen excretion indicative of the conversion of protein into carbohydrate.
5. The occurrence of acetone and B-hydroxybutyric acid in the blood and urine owing to faulty fat metabolism.

The injection of insulin will relieve these symptoms and restore to normalcy the distorted metabolic pattern.

The main effect of insulin is to increase the utilization of glucose by most today tissues. In the whole animal the most important overall effect of insulin quantitatively is to increase the rate of glycogen formation and oxidation of carbohydrate in muscles. The primary effect of insulin in increasing glucose utilization is to transport glucose across the cell membrane into the cell. Insulin has been shown to increase the penetration into cells of many monosaccharide. Insulin appeared to promote the entry into cells of such sugars which possess the same chemical configuration at carbon atoms 1,2 and 3 does D-glucose, The insulin effect has been demonstrated in cats, dogs, rats, rabbits and man and on diaphragm, erythrocytes, heart and skeletal muscles, but not in brain. It is of interest that insulin is not necessary in the metabolism of brain tissue.

In relation to the action of insulin in increasing the passage of sugar into the cells the concept of a membrane carrier system as the functional mechanism governing the transport of glucose and other sugars is considered the most valid. According to the theory, the sugar on the outside of the cell membrane is postulated to form a complex with a "Carrier" substance in the membrane. The complex then moves across the membrane. Within the cell the "sugar-carrier" complex dissociates and releases free sugar. The free carrier may both combine with free sugar in the cell and transport it to the extracellular fluid; of may move back to the outside membrane to complex with free sugar and transport it into the cell.

Nothing is known about the chemical nature of the carrier. Presumably it may be a lipid character which combines reversibly with sugar molecules; the complex formed possessing membrane solubility.

As primary action of insulin on Carbohydrate metabolism appears to reduce blood sugar concentration by greatly enhancing the rate of entrance of sugar to the intracellular compartment, and considering the carrier theory of sugar transport to be valid, the action of insulin may be implied to occur upon some component of the carrier system.

Insulin particularly facilitates the transport of glucose into the skeletal and cardiac muscles and possibly into the adipose tissues. In the transport of glucose into the tissues such as liver, brain, kidney, intestinal mucosa etc. insulin does not appear no appear to be involved. These tissues offer much less resistance to glucose penetration.

Within the cell, glucose enters the metabolic stream. At the initial stage the metabolism of glucose generally involves phosphorylation by ATP to the glucose-6-phosphate in a reaction catalyzed by the glucokinase enzyme in the hexokinase reaction.

Glucose + ATP α glucose – 6 Phosphate + A.D.P.

The activation of glucokinase and glucose phosphorylation is partly effected by insulin and opposes the inhibitory influence of the pituitary growth hormone and adrenal steroids in transformation of glucose to glucose-6-phosphate. Insulin has been found to increase glucagon synthetase activity in muscle.

The Role of Insulin in Fat Metabolism

Insulin is indirectly responsible for the conversion of glucose to fat I the liver and in the adipose tissue. The essential steps of biosynthesis of fat from glucose are conversion of glucose to fatty acid and combination of fatty acids with glycerol to from neutral glycerides which are stored largely in the adipose tissue in the liver.

In this process, pyruvic acid formed from glucose reacts with coenzyme A to form acetyl coenzyme A which is reduced to fatty acid at the expense of hydrogen, derived from reduced nicotinamide adenine dinucleotide phosphate or $NADPH_2$. Reduction of NADP to $NADPH_2$ is effected by hydrogen derived from the glucose-6-phosphate. Hydrogen of $NADPH_2$ receives oxygen from glucose. Fatty acids thus formed combine with glycerol to from neutral fat. Again glycerol is formed from the glucose-6-phosphate. Thus the storage of fat is also influenced by insulin.

The Role of Insulin in Protein Metabolism

Investigation indicate that insulin stimulates protein synthesis by increasing transport of amino acids into cells and at the same time stimulation nucleic acids, particularly RNA and the messenger RNA, which is especially involved in protein synthesis. Insulin has been shown to stimulate RNA synthesis in many cases. Thus it appears that insulin primarily promotes protein synthesis through its effect upon the RNA synthesis. Moreover the secondary effect of insulin is in increasing carbohydrate metabolism and formation of ATP supply energy for protein synthesis.

In the diabetic patient, glucose can be fully utilized for the production of energy and the deficiency of glucose is compensated by the utilization of fat and protein. Breakdown of protein serves as a second source of energy. In the diabetic patient, tissue catabolism and increased excretion of nitrogen produce a negative nitrogen balance.

Thus it has been established that the anabolic and catabolic processes of the three essential food constituents (Carbohydrate, fat and protein) are insulin dependent.

The role of insulin, therefore, appears to be as follows:

1. It facilitates the passage of glucose through cell barriers into the cells.
2. It effects the phosphorylation of glucose.
3. It also plays a role in oxidative phosphorylation.
4. It is essential to lipid and protein catabolism and anabolism.

Diabetes Mellitus

It was referred to as honey urine and melting away of the flesh in the urine. In 1776, Matthew Dobson discovered a chemical method of proof that the sweetness of diabetic urine was due to the presence of glucose. The disease is characterized by clinical symptoms resulting from real or apparent insufficient pancreatic secretion of insulin or possibly over-abundance of some insulin inhibiting factors such as hypersecretion of anterior pituitary or adrenal gland. The result is an elevated blood sugar level (normal range is 80-120mg/100ml of whole blood) which results in pronounced glycosuria tending towards Ketosis.

PLANTS FOR ANTIDIABETIC

1. *Abroma augusta* Linn. (Sterculiaceae)

Common names
Eng.: Perennial Indian Hemp, Devil's Cotton
Hindi & Beng.: Ulat Kambal.

Distribution

A shrub, found throughout the hot parts of India & Pakistan- Cultivated mainly in Uttar Pradesh and Assam.

Parts used: Leaves

Pharmacological activities

Leaves used for uterine disorders, rheumatic pains, sinusitis and diabetes.

Chemical constituents

Taraxerylacetate, taraxerol, and B-sitosterol are found in leaves. Choline, betaine and alkaloid B-sitosterol and stigmasterol are obtained from roots. B-Sitosterol and friedelin isolated; octacosane-1, 28-diol from heartwood.

2. *Acacia melanoxylon* R. Br. (Fabaceae)

Common name
Eng.: Australian Black Wood.

Distribution

Introduced around 1840 in the Nilgiris and has become naturalized.

Parts Used: Roots.

Pharmacological activities

Seeds produced marked hypoglycemia and hypocholesterolemic effects in normal as well as in alloxan diabetic albino rats.

Chemical constituents

Quercetin-3-galactoside (hyperin or hyperoside) m.p. 227-30^0 form flowers was isolated. Stigmast-7-enola and α-spinasterol from heartwood was isolated.

3. *Acacia modesta* Wall. (Fabaceae)

Common name: Punjab: Phulai.

Distribution

It is found in Punjab, sub-Himalayan tract and outer Himalayas, ascending up to 4,000 ft.

Parts used: Seeds.

Pharmacological activities

Seed diet exhibited hypoglycemic effect in normal rats.

Chemical constituents: It contains gum.

4. *Acacia nilotica* Linn. (Fabaceae)

Common Names
Eng.: Babul bark
Hindi: Babool
Beng.: Babula

Distribution

A moderate sized spiny evergreen tree-common all over India, in dry and sandy localities, plentiful in Western Peninsula, the Deccan and Coromandal coast.

Parts used: Seeds.

Pharmacological activities

Seed diet reduced blood sugar, level in normal rats.

Chemical constituents

Bark contains a large quantity of tannin, pods contain also tannin. Gum contains Arabic acid combined with calcium, magnesium and potassium and small amount of malic acid. A new arabinobiose-2-O-B-L-arabinose along with known 3-O-B-L-arabinopyranosyl-L-arabinose from gum was isolated.

5. *Aconitum ferox* Wall. (Ranunculaceae)

Common names
Eng.: Indian Aconite, Monkshood
Hindi: Mithazahar
Beng: Kathbish or Mithavish

Distribution

Distributed at sub-Alpine regions of the Himalayas, rat ward of Kumaon, Nepal, Kashmir and Sikkim.

Parts used: Roots.

Pharmacological activities

Roots has diaphoretic, diuretic, anti-periodic antipyretic and anti-diabetic action in very small doses. For internal administration of the tincture of the roots must be used with great caution on account of the high toxicity. It should not be used in the case when heart disease is present.

Chemical constituents

The tuberous root contains crystalline toxic alkaloid called napelline or Pseudo-aconitine, similar to aconitine, and small quantity of aconitine, picro-aconine, benzyl aconine and homo napelline.

6. *Adhatoda vasica* Nees. (Acanthaceae)

Common names
Hindi: Arusha
Beng.: Basak or Vasaka.

Distribution

It is sub-herbaceous bush, found throughout the year in plains and sub-Himalayan tracts in India, ascending up to 1200 metre, flowers during February-March and also at the end of rainy season.

Parts used: Leaves.

Pharmacological activities

Leaves are commonly used in bronchial troubles. Oral administration of the leaves reduced blood sugar in rabbits for a short period of time.

Chemical constituents

The chief principles of leaves are alkaloids vasicine (MW188, $C_{11}H_{12}N_2O$), Vasicinone and vasicinol. Beside these B-sitosterol, tritriacontane, vasinine, an essential oil and a resin are also obtained from leaves. Beside these new quinazoline alkaloids such as adhatodine, anisotine, vasicoline and vasicolinone were isolated.

7. *Adiantum capillus-veneris* Linn. (Polypodiaceae)

Common names
Eng.: Maidenhair Fern
Hindi & Kan.: Hansraj, mubaraka, pursha.

Distribution

Chiefly obtained in the Punjab bazaar and in some parts of South India.

Parts used: Whole plant.

Pharmacological activities

The fern is used as an expectorant and tonic. Whole plant produced hypoglycemic activity in normal rabbits.

Chemical constituents

The fern contains 3a, 4a-epoxy-filicane, 21-hydroxy-adiantone, and adiantone. It also contains an essential oil.

8. *Adiantum incisum* Forsk. (Polypodiaceae)

Common names
Hindi: Morshikha (Talmurga)
Beng.: Mayurshikha.

Distribution

A Himalayan from, also cultivated in garden as ornamental plants.

Parts used: Roots.

Pharmacological activities

Used in hemicrania and diabetes.

Chemical constituents

It contains adiantone, iso-adiantone, fernene, hentriacontane, hentriacontanone-16, and b-sitosterol.

9. *Aegle marmelos* Coor. (Rutaceae)

Common names
Eng.: Bael Tree

Hindi: Beng & Mar, Bel.

Distribution

A tree, attaining a height of 12 m growing wild and also cultivated throughout the country, particularly in the dry regions.

Parts used: Leaves and Roots.

Pharmacological activities

Marmelosin of the fruit acts as a laxative and diuretic. Unripe and half ripe fruits are stomachic, digestive and used in diarrhea and dysentery.

Leaves and roots exhibited hypoglycemic activity in normal rabbits.

Chemical constituents

Umbelliferone, skimmianine, marmin, β-sitosterol lapel and Y-sitosterol from immature bark and roots. Leaves contain tannin phlobatannin aegeline, flavon-3-ols, leucoanthocyanidin, anthocyanins. Fresh leaves yield on distillation a yellowish-green volatile oil with a peculiar odour. Four alkaloids, 0-(3,3-dimethyl-allyl)-half ordinol, N-2-ethoxy-2-(4-methoxy-phenyl) ethyl cinnamamide, N-2-methoxy-2-[4-(3,3-dimethyl-allyoxy)phenyl] ethyl cinnamamide and N-2-nethoxy-2-(4-methoxyphenyl) ethyl cinnamamide, isolated from leaves. In fruit pulp in addition with normal substance contains a body named marmelosin, which is considered to be one of the most important active principals of the fruit.

10. *Albizia stipulate Sensu* Baker. (Mimosaceae)

Common names
Hindi: Siran
Beng: Chakua.

Distribution

Distributed in the sub-Himalayan tracts, Assam, Bengal, S. India and the Andamans, Extensively cultivated for shade in the coffee and tea gardens.

Parts used: Seeds.

Pharmacological activities

Marked hypoglycemic activity was produced in normal albino rats by the seed diet but no hypoglycemic effect was found in alloxan-diabetic albino rats.

Chemical constituents: It yields gum.

11. *Allium cepa* Linn. (Liliaceae)

Common names
Eng.: Onion, Hindi: Piyaj, Beng.: Pyanj.

Distribution

Cultivated all over India.

Parts used: Bulbs.

Pharmacological activities

Onion has diuretic, stimulating and expectorant actions, it is also used as antiflatulent.

Allyl propyl disulphide from the bulbs reduced blood sugar level of alloxan-diabetic rabbits.

Significant hypoglycemic effect was produced in mice by onion oil and synthetic dipropyl- disulphide oxide.

Oral administration of allicin (diallyl disulphide oxide) exhibited significant reduction of the blood sugar levels in alloxan diabetic rabbits.

Hypoglycemic activity was produced by the ether extract of juice expressed onion bulbs in diabetic rabbits.

Oral administration of allyl propyl disulphide to the human volunteers caused significant fall in blood glucose level.

Petroleum ether extract of bulbs exhibited fall of blood sugar in rabbits.

Hypoglycemic activity was produced in rabbits by the light petroleum extract of dried onion.

Onion tops caused the fall of fasting blood sugar level in alloxan and adrenaline-diabetic rats.

Total extract of dried onion bulb produced hypoglycemic activity in rats and rabbits.

Petroleum ether and chloroform extracts caused a significant fall in blood sugar at the time of glucose tolerance lost.

Ingestion of onion bulb juice caused the lowering of the blood sugar level in diabetes mellitus patients.

Two fractions, fraction A and B from the onion bulb juice produced lowering of blood sugar level in rabbits.

Anti-diabetic activity was produced by onion bulbs extracts in rabbits.

On oral application of hypoglycemic fraction from the onion bulbs to alloxan diabetic rabbits increased glucose tolerance. The juiced expressed residue of the bulb showed hypoglycemic activity in diabetic patients.

The non-dialyzable fraction from the lipid free juice of the onion bulb exhibited hypoglycemic activity in rabbits.

Insulin like activity was observed in onion bulbs extracts, in rabbits. Onion bulbs extract produced lowering of the blood sugar level in human volunteers during the glucose tolerance test, but no hypoglycemic effect was observed on fasting blood sugar level.

Hypoglycemic activity was observed in the extracts of green sprouting taps, roots and bulbs of onion in rabbits.

Chemical constituents

Essential oil of whole plant is 0.05%. Chief constituent of include oil is allyl-propyl-L-cysteine sulphoxide isolated as N-2,4-dinitrophenyl derived from bulb. Propyl sulfonic acid was obtained from bulb.

12. *Allium sativum* Linn. (Liliaceae)

Common names
Eng.: Garlic
Hindi & Guj.: Lasan
Beng. & Mar.: Lasun.

Distribution

Cultivated all over India.

Parts used: Bulbs.

Pharmacological activities

Garlic juice is applied in skin troubles and used as ear drops. This juice is also used in dyspepsia and flatulence. Garlic is considered stimulant, expectorant and diuretic,

Ethyl ether extract of dried garlic bulb powder exhibited hypoglycemic activity in glucose feeding normal fasting rabbits.

Hypoglycemic activity was produced by the garlic bulb in alloxan diabetic rabbits.

Hyperglycemic effect of glucose feeding in rabbits was controlled by the garlic juice.

Chemical Constituents

See Hepatoprotective chapter.

13. ***Aloe barbadensis*** **Mill. (Liliaceae)**

Common names
Eng.: Curacao Aloe, Barbados Aloe
Hindi: Gheekanvar
Beng.: Ghrita-kumari

Distribution

It is xerophytic, arborescent or herbaceous, the fleshy and strongly cuticularised leaves usually prickly at the margin, and arranged in dense rosettes. It is naturalized in India. It is planted in many Indian gardens and available all over India.

Parts used: Leaves.

Pharmacological activities

Fresh juice is used as cathartic. It is also very useful in x-ray burn and any other radiation burns.

Semi-transparent, amorphous solid of the fresh leaves produced hypoglycemic effect in normal albino rabbits when administered intravenously and significant hypoglycemic effect was also found in alloxan diabetic rabbits.

Chemical constituents

Leaf latex contains aloin, isobarbaloin, emodin, aloe-emodin, B-barbalin.

14. ***Alpinia galanga L.*** **Willd. (Zingiberaceae)**

Common names
Eng.: The Greater Galangal
Hindi & Beng: Kulanjan.

Distribution

Found in South India and Bengal.

Parts used: Rhizomes.

Pharmacological activities

Dried rhizomes used as carminative and stimulant.

Essential oil from rhizomes used in perfumery. It is also used in dyspepsia, fever and diabetes mellitus.

Chemical constituents

Compheride, galangin and alpinin. From the green rhizome a pale yellow volatile oil with a pleasant odour can be obtained by distillation. This volatile oil contains methyl cinnamate, cincole, camphor. Seeds contain caryophyllene oxide, caryophyllenol,1′-acetoxychavicol acetate, 1′-acetoxyeugenol acetate, 1′-acetoxyeugenol acetate pentadecane and 7-heptadecene.

15. ***Anacardium occidentale*** **Linn. (Anacardiaceae)**

Common names
Eng.: Cashew Nut Tree
Hindi & Mar.: Kaju
Beng.: Hijli-badam

Distribution

An evergreen tree generally cultivated and naturalized in the hotter parts of India especially near sea. Flowers are yellow with pink stripes.

Parts used: Leaves and barks.

Pharmacological activities

Oral application of the tincture or extract of the bark produced hypoglycemic activity in normal individuals.

Leaves exhibited hypoglycemic activity in albino rats. Intravenous administration of bark extract reduced the hyperglycemia of alloxan diabetes dogs and rats.

Chemical constituents

A gum containing bassorin, partially soluble in water exudes from the bark.

Leaves contain p-hydroxy benzoic, protocatechuic, gentisic, gallic acids, and glycosides of kaempferol and quercitol.

16. *Arctium lappa* Linn. (Asteraceae)

Common names
Eng.: Burdock.

Distribution

Found at W. Himalayas from Kashmir to Simla.

Parts used: Leaves

Pharmacological activities

Extracts caused sharp long-lasting hypoglycemic effect with increase in carbohydrate tolerance in rats.

Chemical constituents

Eremophilence, fukinone, betasotolone, fukinanolide b-eudesmol, taraxasterol its acetate and palmitate and dehydro-fukinone and aerator (8a-hydroxyeudesmol) obtained from leaves.

17. *Areca catechu* Linn. (Arecaceae)

Common names
Eng: Areca Nut
Hindi & Beng.: Supari

Distribution

Cultivated throughout tropical India. It flourished in dry plateau of Mysore, Canara, Malabar, Assam etc. Area nuts are most commonly used as a masticatory.

Pharmacological activities

Alkaloid fraction exhibited hypoglycemic activity

Chemical constituents

See Antifertility chapter.

18. *Avena sativa* Linn. (Cyperaceae)

Common names
Eng.: Oat, Common Oat
Hindi & Beng.: Jai.

Distribution

Annual or perennial grasses. Cultivated to a limited extent in W. Himalayas, Sikkim and N. West Bengal. Available in Indian bazaar and many other countries.

Parts used: Seeds.

Pharmacological activities: Seed is used in Diabetes.

Chemical Constituents: See Anti-fertility chapter.

19. *Azadirachta indica* A. Juss (Meliaceae)

Common names
Eng.: Margosa Tree
Hindi & Beng.: Nim; Neem.

Distribution

A large tree with rough bark. Native to India, grown all over India. Grows wild in the dry forests of the Deccan.

Parts used: Leaves.

Pharmacological activities: The losses in body weight of rats caused by prolong administration of anterior pituitary extract was prevented by "Tribang shila" a composite drug containing neem leaves. The hyperglycemic response of anterior pituitary extract in rats, reduced significantly by "Tribang Shila" a composite drug containing neem leaves.

Aqueous extract of leaves produced significant reduction of the blood sugar level in diabetic dogs.

20. *Bauhinia semla* Wunder Linn. (Mimosaceae)

Common name
Hindi: Semla.

Parts used: Seeds.

Pharmacological activities

Hypoglycemic as well as hypocholesterolemic activities were produced in normal and alloxan-diabetic albino rats by seeds.

Chemical constituents

Bark contains quercetin-3-0-/b-D-glucoside and rutin.

21. *Benincasa hispida* Cogn. (Cucurbitaceae)

Common names
Eng.: Ash Gourd
Hindi: Petha
Beng.: Chalkumra.

Distribution

A large climber and annual plant. Cultivated through out India; fruits used as vegetable. Flowers are large yellow.

Parts used: Fruits.

Pharmacological activities

An ayurvedic medicine. "Kushmanda lehyam" is used for diabetes.

Chemical constituents

β-Sitosterol, lupeol, n-triacontanol, mannitol, arginine, aspatic acid, protine, hydroxyprotine, isoleucine cysteine, glutamic acid, glucose rhamnose found in fruits.

22. *Bougainvillea spectabilis* Willd. (Nyctaginaceae)

Distribution

A shrub, flowers are small, usually enclosed by large, purple, of white or orange coloured bracts, commonly grown in gardens.

Parts used: Leaves.

Pharmacological activities

Hypoglycemic action of the leaf-juice was established in alloxan induced diabetic rabbits on oral administration. Moreover leaf juice also increased liver and muscle glycogen in alloxan diabetic rats on repeated administration.

23. *Brassica oleracea* Linn. (Brassicaceae)

Common names
Eng.: Cabbage
Hindi: Band-gobhi, patagobhi
Beng.: Bandhakopi.

Distribution

Cabbage is a commonly used vegetable, like cauliflower and available throughout India.

Parts used: Leave

Pharmacological activities

No extract was able to be prepared from cabbage containing hypoglycemic activity.

No significant hypoglycemic effect was produced in morning blood sugar level on application of cabbage extracts orally and subcutaneously.

One of the two fractions of cabbage extract caused an increase in blood sugar, associated with glycosuria and decrease of the liver glycogen content but the other fraction exhibited hypoglycemic effect in normal rabbits and caused apparent replacement of insulin in the depancreatized dog.

Chemical constituents

Leaves contain quercetin, isorhamnetin, and 3-sophoroside-7-glucosides of kaempferol. More over cyanidin-3-p-coumaryl-sophoroside-5-glucoside, cyanidin-5-glucoside, cyanidin-3-(disinapyl)-sophoroside-5-glucoside, cyanidin-3-sophoroside-5-glucoside etc.

24. *Casearia esculenta* Roxb. (Samydaceae)

Common names
Mar.: Mori, Kulkulta
Tam.: Kottargovai.

Distribution

Found at W. Peninsula.

Parts used: Roots

Pharmacological activities

Roots used in diabetes and piles. The active hypoglycemic principles a resin fraction and a crystalline compound (m.p. 1820) have been isolated from the roots.

The water soluble crystalline fraction and the resin part of the root produced hypoglycemic effect.

Significant lowering of blood sugar was observed in rats by the decoction of the roots, and glucose tolerance was favorably influenced on chronic administration. Alcoholic extract of the roots caused hypoglycemic effect in albino rats. Aqueous and alcoholic extracts influenced favourably the glucose tolerance.

Chemical constituents

Leucopelargonidin, m.p. 220^0C and two sterols of m.p. 120^0 and 123^0 C obtained from roots.

25. *Cass auriculata* Linn. (Mimosaceae)

Common names
Eng.: Tanner's Cassia, Avaram
Hindi: Tarwar
Guj.: Awal
Mar.: Tarwad.

Distribution

Found in Madhya Pradesh and W. Peninsula. Cultivated elsewhere.

Parts used

Seeds, Leaves and flowers.

Pharmacological activities

Hypoglycemic effect was observed on application of water extract of the seeds in normal rabbits and alloxan rabbits and dogs.

Chemical constituents

B- Sitosterol and kaempferol were isolated from flowers. Leaves contain saturated higher fatty ketoalcohols and emodin.

4,5,7-trihydroxyflavan-3, 4-diol also known as goratensidine and (-) auricul-acacidin were detected.

26. *Cassia fistula* Linn. (Fabaceae)

Common names
Eng.: Indian Laburnum
Beng.: Amaltas, Kalkasunda
Hindi: Bandarlathi.

Distribution

A moderate sized deciduous tree, common throughout India, as wild and cultivated. Fruits are collected when ripe.

Parts used: Seeds.

Pharmacological activities

Seed diet produced marked hypoglycemic effect on normal albino rats but caused no hypoglycemic effect on alloxan diabetic albino rats.

Chemical Constituents

See Anti-inflammatory chapter.

27. *Cassia occidentalis* Linn. (Fabaceae)

Common names
Eng.: Negro coffee
Hindi: Kasondi
Beng.: Barokalkesunda.

Distribution

A common weed found from the Himalayas' to the West Bengal and South India.

Parts used: Roots and Leaves.

Pharmacological activities

It has same pharmacological actions as *C. sophera* as per Sanskrit writers.

Chemical constituents

Roots contain anthraquinones, a phytosterol, a hydroxyanthraquinone, m.p. $128^0$1,8-dihydroxanthraquinone, emodin, quercetin, an anthraquinone m.p. 179^0 and a substance similar to rhein and leaves contain dianthronic heteroside.

Chrysophanol and emodol from young roots and C-flavonosides of apigenin from pericarp were isolated.

28. *Cassia sophera* Linn. (Fabaceae)

Common names
Eng.: Senna Esculenta
Hindi: Kasauns
Beng: Chhoto-Kalkesunde

Distribution

A shrub 2-3 m high annual or perennial, found throughout the tropical parts of India.

Parts used: Seeds and bark.

Pharmacological activities

Infusions of bark or powdered seeds with Honey are given in diabetes.

Chemical constituents

Emodin and chrysophanic acid flowers contain rhamnetin-3-O-B-D-glucoside and chrysophanol.

29. ***Catharanthus roseus*** **G. Don. (Apocynaceae)**

Common names
Hindi: Sadabahar
Beng.: Nayantara

Parts used: Whole plant.

Pharnacological activities

It is used in diabetes. Ether and chloroform soluble alkaloid fraction of the whole plant caused marked anti-diuretic activity in rats. It produced hypoglycemic activity. Main alkaloids fraction from the plant reduced blood sugar level in rats. No significant hypoglycemic effect was produced by the water extract of leaves in normal and diabetic rabbits and dogs.

Chemical constituents

Four alkaloids possessing antibacterial activities from leaves and two glycosidal principles and ursolic acid, leurosine, isoleurosine, previniue, mitraphyline, lochnevin, personine.

30. ***Cichorium intybus*** **Linn. (Asteraceae)**

Common names
Eng.: Chicory, Wild endive
Hindi: Kasani, Kasni.

Distribution

The chicory is native of Europe, found in India, cultivated at elsewhere.

Parts used: Leaves.

Pharmacological activities

Leaves produced no hypoglycemic activity in alloxan diabetic rats. Water extract of leaves produced hypoglycemic activity in rabbits (unpublished work).

Chemical constituents

Aerial parts contain quercitrin, apigenin, hyperin, luteolin-7-B-D-glucopyranoside, caffeic, dicaffeoyltar acids, apigenin-7-O-L-arabinoside, chlorogen and neochlorogenic and inflorescence contains umbelliferone, 6,7-dihydroxy coumarin, cichoriin and esculin.

31. ***Citrus aurantium*** **Linn. (Rutaceae)**

Common names
Eng.: Common orange
Hindi: Narengi
Beng.: Kamlaneboo.

Distribution

Found in northern India, and different varieties are distributed chiefly in the warmer moist regions of India such as Assam.

Parts used: Fruits.

Pharmacological activities

The laevulose found in orange is beneficial in diabetes.

Chemical constituents

Rind contains volatile oil, hesperidin,

neohesperidin etc. Orange fruit contains laevulose.

32. *Clerodendrum phlomidis* Linn. (Verbenaceae)

Common names
Hindi, Guj. & Mar.: Arni.

Distribution

It is found at the Gangetic valley.

Parts used: Whole plant.

Pharmacological activities

Roots are aromatic and astringent. Decoction and alcoholic extract of the crude drug, "Arni" is useful for diabetic mellitus patients. Alcoholic extract of the whole plant exhibited hypoglycemic effect in normal and diabetic mellitus patients.

Chemical constituents

Scutellarin and pectolinarigenin (4',6-dimethyl scutellarin) obtained from leaves. Leaves also contain B and Y- sitosterols, a monoglucoside m.p. 213^0ceryl alcohol, palmitic and cerotic acids, and an steros m.p. 155^0. Stem contains B-sitosterol-B-D-glucoside, ceryl alcohol and B-sitosterol.

33. *Coccinia indica* Wight. & Arn. (Cucurbitaceae)

Common names
Eng.: Lvy Gourd
Hindi: Kundru

Parts used: Roots.

Pharmacological activities

Water soluble fraction of alcoholic extract of fat free roots exhibited hypoglycemic effect in alloxan-diabetic rabbits. Alcoholic extract of roots produced hypoglycemic activity in alloxanized rats and rabbits. Aqueous and ethanolic extracts of the defatted powdered roots significantly reduced blood sugar level in alloxan-diabetic rabbits comparable to tolbutamide.

No antidiabetic action was observed either in blood sugar or in urine of patients suffering from glycosuria by the fresh juice of leaves, stem and roots. No hypoglycemic activity was produced by the amylolytic enzyme, a hormone and an alkaloid isolated from the whole plant in rabbits. Acute hyperglycemic condition, produced by the anterior pituitary extract in rats was inhibited by the alcoholic extract of fruit. Hypoglycemic activity for a short duration and activity on glucose tolerance test was observed in normal guineapigs caused by a quaternary base isolated from this herb. A moderate lowering of blood sugar was also produced n alloxan diabetic rats.

Chemical constituents

B-amyrin, lupeol, and cucurbitacin B are found in fruits. Aerial parts contain cephalandrol, m.p. 81^0 tritriacontane, B-sitosterol cephalandrine A and cephalandrine B. Stigmast-7-en-3-one m.p. 154^0 isolated from roots.

34. *Cryptostegia grandiflora* R.Br. (Asclepiadaceae)

Common names
Mar.: Vilayati vakundi
Tam.: Plalai.

Distribution

Cultivated mostly as a hedge plant in gardens.

Parts used: Aerial parts.

Pharmacological activities

Alcoholic extract of the aerial parts produced significant hypoglycemic effect on normal rabbits but no reduction of blood sugar occurred in alloxan diabetic rabbits. It was hepatotoxic.

35. *Cyamopsis tetragonolobas* Linn. (Fabaceae)

Common names
Eng.: Cluster bean
Hindi: Govar, Guar
Tam.: Kothaveray.

Distribution

It is found from the plains of the Himalayas to W. Peninsula. Widely cultivated, pods used as vegetable.

Parts used: Seeds.

Pharmacological activities

Significant reduction of blood sugar levels were observed in diabetic and non-diabetic adults after taking meals containing seeds gum. Blood glucose levels remain unaltered with gum and the post prandial blood glucose curve was flattened. Excretion of glucose through urine decreased significantly in diabetic patients.

The rise of blood glucose level reduced significantly after taking meal containing seed gum. No significant difference was found between the mean blood sugar levels of healthy volunteers after ingestion of gelled or non-gelled seed gum. An insignificant effect on post-prandial glucose level was observed in human beings after ingestion of seed gum.

36. *Cynara scolymus* Linn. (Asteraceae)

Common names
Eng.: Burr Artichoke
Hindi & Beng. Hathichoke

Distribution

Cultivated to a limited extent throughout India. Young flower heads are eaten as a vegetable. Leaves are bitter.

Parts used: Whole Plant.

Pharmacological activities

Flower head produced no hypoglycemic effect on normal. Hypoglycemic activity was observed when a dialyzed extract of the artichoke was injected into rabbits.

Chemical constituents

Leaves contain cynarin m.p. 235^0 cynaropicrin, cynarolide m.p. 126^0, caffeic acid, 7-Brutinoside, 1,3,4,5-caffeoyl, luteolin-7-B-D-glucoside, and cynarotrioside, m.p. 274^0 Cynarogenin, taraxasterol, stigmasterol and B-sitosterol are obtained from receptacles.

37. *Daucus carota* Linn. (Apiaceae)

Common names
Eng.: Carrot
Hindi: Gajar
Beng.: Gajar.

Distribution

The carrot is cultivated throughout the greater part of India. Roots are edible.

Parts used: Roots.

Pharmacological activities

Marked lowering of blood sugar was produced without any toxicity, by an amorphous yellow, colored fraction of petroleum ether extract of the dried tuberous roots, in human beings, rabbits and dogs.

Chemical constituents

See Antifertility chapter.

38. *Dolichos biflorus* Linn. (Leguminosae)

Common names
Eng .: Horse Gram
Hindi: Kulthi
Beng.: Kurtikalai.

Distribution

It is found throughout the greater part of India. An important crop-plant is southwards Maharashtra.

Parts used: Seeds.

Pharmacological activities

Hypoglycemic as well as hypocholesterolemic effect were caused when seeds diet was fed to normal rats.

39. *Dolichos lablab* Linn. (Fabaceae)

Common names
Eng.: Indian Butter Bean, Lablab Bean
Hindi: Sem
Beng.: Shim

Distribution

A climbing herb, cultivated throughout India. Pods are eaten as vegetable. Seeds are also used as food.

Parts used: Pods.

Pharmacological activities

Reduction of fasting blood sugar level in alloxan diabetic rats, as well as antagonism of the adrenaline induced hyperglycemic response was produced by the green pods.

Hyperglycemic effect due to the administration of dextrose with adrenaline, fasting blood sugar level was reduced in experimental animals, by the green pods. Green pods duet also exhibited antiglycosuria, antiacetonuria and antiacetonemia activities in diabetic patients.

Chemical constituents

L-Pipecolic acid was obtained from pods.

40. *Emblica officinalis* Gaertn. (Euphorbiaceae)

Common names
Eng.: Emblic Myrobalan
Hindi: Amla, amlika
Beng.: Dhatri, amlaki

Distribution

A small or medium sized tree, found throughout tropical part of India, ascending up to 133 m, cultivated in gardens and homewards.

Parts used: Seeds.

Pharmacological activities

Infusion of seeds is used in diabetes.

Chemical constituents

In addition with vitamin C fruits contain corilagin, ellagic acid, and Trigalloyl glucose.

41. *Enicostemma littorale* Blume. (Gentianaceae)

Common name
Hindi: Chhota-chirayata

Distribution

It is found throughout India-flowers are white in auxiliary cluster.

Parts used: Whole plants.

Pharmacological activities

"Tribang shila" composite drug, contains E. *littorale* extract, inhibited significantly the hyperglycemic activity of anterior pituitary extract in rats. Fresh juice of whole plant exhibited lowering of fasting blood sugar level in diabetic patients without any toxic effect.

Oral application "Tribang shila" prevented the loss in body weight in rats caused by the prolonged application of anterior pituitary extract.

Chemical constituents

It contains gentiocrucine, a mixture of three compounds and a monoterpene alkaloid-enicoflavine.

42. *Ensete superbum* Roxb. (Musaceae)

Common names
Mar.: Chowani.

Distribution

Distributed from Bombay to Western Ghats to Travancore hills and ravine slopes and in Assam.

Parts used: Seeds.

Pharmacological activities

Hypoglycemic action was produced in experimental animal without any harmful effect by a fraction of seeds extract.

Chemical constituents

In the seeds three constituents were detected.

43. *Eriodendron anfractuosum* DC. (Bombacaceae)

Common names
Eng.: Kapok Tree, White Silk, Cotton Tree
Hindi: Safedsimul
Beng.: Safet shimul.

Distribution

The tree is found in the hotter parts of India, Ceylon etc.

Parts used: Seeds.

Pharmacological activities

Extract used in diabetes.

Chemical constituents

Seeds contain fixed oil which is triglycerides of palmitic, linoleic, and oleic acids. Gum-produced by tree contains gallic and tannic acids.

44. *Erythrina indica* Lam. (Fabaceae)

Common names
Eng.: Indian coral tree
Hindi: Dadap
Beng.: Palita-mandar, Palidhar.

Distribution

A common tree of Bengal, Southern India and many other parts of India, generally planted for shade and support.

Parts used: Root bark.

Pharmacological activities

A decoction of the root bark with a dose of 'Vasant Kusumakar' Rasa is said to reduce the quality as well as sugar of urine on an application on every morning in diabetes cases.

Chemical constituents

Bark contains B-sitosterol, y-sitosterol, m.p. 148^0, S-sitosterol, m.p. 147^0, docosyl alcohol and three other substances having m.p. 135^0, 126^0 and 142^0.

45. *Eucalyptus citriodora* Hook. (Myrtaceae)

Distribution

It is an aromatic evergreen tree. Grows in North India, Kerala and Nilgiri. Leaves yield essential oil.

Parts used: Leaves.

Pharmacological activities

Leaves extract, on oral administration, caused lowering of blood sugar in normal and alloxan-diabetic rabbit and it also flattened the blood sugar curve in glucose tolerance test action is considered due to myrtillin in extract.

Chemical constituents

Citronellal, geraniol isoluegol, linalool, cineole and many other essential oils were detected in leaves.

46. *Ficus bengalensis* Linn. (Moraceae)

Common names
Eng.: Banyan Tree
Hindi: Bar, Bargad
Beng.: Bot, Bar.

Distribution

The banyan tree is a large branching tree with numerous aerial roots occurring all over the plain part of India. Planted for shade.

Pharmacological activities

Hypoglycemic activity was observed in normal and moderately diabetic rabbits produced by the bengaranoside isolated from the bark. Ethanolic extract of bark exhibited hypoglycemic activity in rabbits. Flavonoids A, B & C from bark produced hypoglycemic effect in normal.

No appreciable difference in blood sugar levels was caused in diabetic rabbits by the decoction of bark. Hypoglycemic action was produced by the ethanolic extract of bark in normal male albino rats. Aqueous extract of the bark caused no significant hypoglycemic action on normal fasting rabbits, but produced moderate lowering of fasting blood sugar levels in alloxan diabetic rabbits. Shilajit and milky sap from the plant produced initial lowering of blood sugar in anterior pituitary extract treated albino rats.

Bark extract delayed glucose absorption significantly in mice but without any toxicity, on oral application of aqueous extract of bark.

Hypoglycemic activity was produced in rats by the glycosidal fraction from the aqueous extract of bark.

Chemical constituents

Quercetin-3-galac to side, rutin, friedelin and B-sitosterol were isolated from leaves, and taraxasterol tiglate from heart wood. Stem bark contains three methyl ethers of leucoanthocyanodins delphinidae-3-O-α-L-rhamnoside, pelargonidin-3-O-α-L-rhamnoside and leucocyanidin-3-O- Beta-D-galactosyl cellobioside and methyl ether of leucoanthocyanidin.

47. *Ficus racemosa* Linn. (Moraceae)

Common names
Eng.: Cluster Fig, Country Fig
Hindi: Gular
Beng.: Jagyadumur

Distribution

A large deciduous tree distributed throughout India, particularly in evergreen forests, moist localities. It is cultivated in village for shade and edible fruits.

Parts used: Bark and stem bark.

Pharmacological activities

Stem bark caused hypoglycemic activity in albino rats.

Bark exhibited hypoglycemic effect in normal rabbits.

Aqueous decoction of the bark produced no effect on blood sugar levels of rabbits.

Chemical constituents

Leaf contains glycoside and bark contains tannin.

48. *Ficus religiosa* Linn. (Moraceae)

Common names
Eng.: Pipal Tree
Hindi: Pipli, papal
Beng.: Aswatha

Distribution

A large perennial tree, mostly planted as road-side tree particularly near temples. It is found all over the plains of India.

Parts used: Bark and roots.

Pharmacological activities

Bark contains tannin used in skin troubles.

Aqueous extract of bark produced hypoglycemic activity in rabbits.

Water extract of root bark exhibited hypoglycemic effect in rabbits.

Sitosterol glucoside of bark decreased blood sugar level of rabbits.

No effect was produced on the blood sugar level of rabbits by the aqueous decoction of bark.

Chemical constituents

Bark contains tannin; B-sitosterol-D-glucoside was isolated.

49. *Glycine max* Merrill. (Fabaceae)

Common names
Eng.: Soyabean, Soya
Hindi: Bhat, Bhatwar
Beng.: Garjkalai

Distribution

Soyabean is a suberect herb, cultivated mainly in Punjab, Himachal Pradesh, Kashmir, Bengal, Bihar and Assam; Seeds are used as a pulse.

Parts used: Seeds.

Pharmacological activities

Seeds produced hypoglycemic activity in normal albino rats.

Chemical constituents

Roots contain flavonoid daidzein and one isoflavone glycoside. Genistin and two galactomonnans were isolated from soyabean hulls.

50. *Gymnema sylvestre* R. Br. (Asclepiadaceae)

Common names
Hindi: Gurmar, Merasingi
Beng.: Merasingi

Distribution

A woody climber, distributed at peninsular India. Flowers are minute greenish-yellow in color, orange spirally in lateral corymbs. Leaves when chewed paralyze the sense of test for sweet and bitter substance for few hours.

Parts used: Leaves.

Pharmacological activities

Alcoholic extract of leaves produced hypoglycemic effect in mild diabetic animals.

An insignificant reduction of blood sugar occurred in normal rats but significant and marked hypoglycemic effect was observed in hyperglycemic animals caused by the alcoholic extract of leaves. Leaf extract produced hypoglycemic activity in rabbits, on oral administration and *in vitro* also.

No effect on blood sugar was observed in rabbits caused by the leaves extracts, gymnemic acid and sodium salt of gymnemic acid. Aqueous and alcoholic extract of leaves, on oral administration, produced hypoglycemic activity in normal human, dogs and diabetic dogs.

Whole plant exhibited hypoglycemic effect on rabbits. Powder and decoction of leaves prevented loss of sugar through urine in human beings.

Hypoglycemic effect was caused in normal and diabetic rabbits and albino rats by the aqueous extract of leaves on oral application.

Leaves and stems parts showed hypoglycemic activity in diabetic human beings but without any toxicity. Aqueous infusion of leaves caused glucose tolerance in albino rats.

Prevention of the loss in body weight in rats caused by the prolonged administration of anterior pituitary extract as well as an increase in the glucose tolerance were caused by the alcoholic extract of leaves.

Leaf powder produced hypoglycemic activity in normal and diabetic subjects of age group of 43 to 68 years.

Chemical constituents

Triterpene-gymnestrogenin and gymnemagenin were detected in leaves. Beside these leaves contain antisaccharin principle, gymnemic acid, characterized as hexahydroxy-olean-12-ene-B-glucuronide m.p. 328^0 and from the hydrocarbon fraction of the leaf extract triacontane and hentriacontane were isolated.

51. *Helicteres isora* Linn. (Sterculiaceae)

Common names
Eng.: East Indian screw tree
Hindi: Marori, Marodphali
Beng.: Atmora.

Distribution

A shrub, found at the Central Peninsula, Central and Western India to Jammu.

Parts used: Root bark.

Pharmacological activities

Juice of root bark or decoction of root bark is applied in diabetes to reduce sugar.

Chemical constituents

An orange-yellow colored crystalline matter, a hydroxyl-carboxylic acid (m.p.a 78-170), saponin, phlobatannins, sugar, and phytosterol were detected in bark.

52. *Hordeum vulgare* Linn. (Gramineae)

Common names
Eng.: Barley
Hindi: Jau, Jav
Beng.: Jab

Distribution

An erect herb. It is cultivated in North-India, Madhya Pradesh, West Bengal and Bihar as a food crop. Barley grain is easily assimilable.

Parts used: Roots.

Pharmacological activities

Hypoglycemic action was exhibited in rabbits by the water soluble fraction of fermented rootless of barley.

Chemical constituents

Two flavor glycosides orientin and orientoside and a luteolin glycoside were isolated. An arabinogalactan-(4-O-methyl-glucurono) xylan was isolated from leaves.

53. *Indigofera arecta* Hochst. (Leguminosae)

Common names
Eng.: Bengal indigo.

Distribution

A shrub, yield a dye-indigo, principle indigo producing plant of Ethiopia.

Parts used: Whole plant.

Pharmacological activities: The plant extract reduced plasma glucose levels of normal fasting rats, and also increased plasma insulin levels.

Chemical constituents

A flavonol glycoside-kaempferitrin (m.p.201-30^0) was isolated from leaves. Kaempferitrin on hydrolysis gives rhamnose and kaemferol. Beside this the plant contains indigotin (approx 1.0%).

54. *Ipomoea nil* Linn.(Convolvulaceae)

Common names
Hindi: Kaladana
Beng.: Hilkalmi, kaladanah.

Distribution

It is a showy plant, and found throughout India.

Parts used: Whole plant.

Pharmacological activities

Extracts reduced blood sugar levels in rats.

Chemical constituents

Arachidic, stearic, palmitic, oleic, linolenic and linoleic acids are obtained from seed oil. Seeds also contain chanoclavine, lylose, xylose, arabinose and galactose.

55. *Kickxia ramosissima* Wall. (Scrophulariaceae)

Common names
Guj.: Kanodi, Bhintgalodi.

Distribution

A perennial herb, found throughout India, on walls, rocky and stony places. Leaves are membranous and flowers are yellow.

Parts used: Aerial part.

Pharmacological activities

Used in diabetes. Hot water extract of the whole aerial parts had hypoglycemic activity in rabbits.

56. *Lagerstroemia speciosa* Pers. L. (Lythraceae)

Common names
Eng.: Queen, Crape Myrtle
Hindi: Jarul
Beng.: Jarul

Distribution

Found from Assam, South wards to Peninsular India, planted in gardens for ornament. Wood is used to a limited extent for furniture, casts, wheels and boxes.

Parts used: Leaves and fruits.

Pharmacological activities

Leaves and ripe fruits produced hypoglycemic activity on oral administration.

Chemical constituents

Alanine, methionine, α-aminobutyric acid and isoleucine were detected. Leaves contain 3,3',4-tri-O-methylellagic acid amyl alcohol, B-sitosterol and tannin-lagertannin, characterized as 3,3-di-O-methyl-4'O-B-D-glucosylellagic acid.

57. *Lupinus albus* Linn. (Leguminosae)

Common names
Eng.: White Lupine,
Hindi: Turmas
Beng.: Turmuz.

Distribution

An erect annual herb with white flower. Grown in gardens. Seeds are used as food as cattle feed after soaking in water for removing toxic alkaloids.

Parts used: Seeds.

Pharmacological activities

A fraction of seed extract caused lowering of blood sugar level in rabbits.

Chemical constituents

Lupanine, sparteine and hydroxylupanine were detected in various parts of the plant. Seedlings and stems are rich source of asparagine. Seeds yield fatty oil. Tiglic, cinnamic and benzoic acid esters of acyloxylupanines and alkaloid angustifoline were isolated from seeds.

58. ***Mangifera indica*** **Linn. (Anacardiaceae)**

Common names
Eng.: Mango
Hindi: Am, Amb
Beng.: Am.

Distribution

A tree, indigenous to India, cultivated widely in many varieties in the plain parts of India for its edible fruits.

Parts used: Leaves.

Pharmacological activities

Dried powder of tender leaves is useful in diabetes.

Chemical constituents

Mangiferolic acid, m.p. 181^0, hydroxymangiferolic acid, isomangiferolic acid, mangiferonic acid, hydroxyl mangiferonic acid m.p. $190^{0,}$ ambolic acid, ambonic acid, amyrin, lupeol, acetates of cycloartenol, homomangiferin, magniferin, m.p. 278^0, protocatechuic acid, friedelim, B-sitosterol, catechin, ellagic acid, gallic acid, m-digallic acid, m-trigallic acid, gallotannin, quercetin, leucocyanidin, disetin and butin were isolated. In leaves galactose, glucose, arabinose rhamnose, xylose, tannin, two phenolic compounds and two flavonoids were detected. From resin triterpenes-oleanolic aldehyde, m.p. 168^0, a diol m.p. 154^0, and an aldehyde, m.p. 166^0 were isolated and in unripe fruits, a polysaccharides, glucan was detected.

59. ***Momordica charantia*** **Linn. (Cucurbitaceae)**

Common names
Eng.: Bitter Gourd
Hindi: Karela
Beng.: Karela.

Ditribution

A climbing herb, found throughout India. Cultivated for fruits, which are used as vegetable.

Parts used: Fruits and seeds.

Pharmacological activities

Fresh juice of fruits produced significant hypoglycemic effect in experimental animals.

"Plant Insulin" the hypoglycemic principle which exhibited a consistent hypoglycemic activity in diabetes mellitus patients was isolated from fruits.

Aqueous extract of fruits suppressed the hyperglycemia in albino rats. Fruits produced hypoglycemic activity in alloxan-diabetic rabbits.

Tablets, prepared from the powder of fruits, or the fresh juice of fruits caused singnificant lowering of blood sugar levels in patients.

On oral administration of fruit juice showed hypoglycemic effect in normal as well as diabetic rabbits. But toxicity was observed, in the animals.

No glycosuria was observed in a diabetic patient, taken fruits along with chlorpropamide.

Fruit juice and the dried extract fruits caused mild hypoglycemic effect in diabetic rabbits.

Unripe fruits showed hypoglycemic action in rabbits.

A non-nitrogenous substance charantin, produced hypoglycemic activity in fasting rabbits was isolated from unripe fruits.

No hypoglycemic effect was produced in rabbits only by the fruit extract alone, rather fruit extract potentialised the hypoglycemic effects of tolbutamide and Jasadbhasma.

Seeds contain a glycol-alkaloid vicine, characterized as 2,6-diaminopyridine-5-B-D-glucopyranoside, showed hypoglycemic activity in fasting albino rats.

No hypoglycemic activity was produced in diabetic and normal human beings by the alcoholic extract of the drug.

Marked hypoglycemic effects were exhibited both by a crude crystalline substance and an infusion of the crude drug. The drug showed toxicity in fish, rats and rabbits.

Chemical constituents

Fruits contain stigmast-5,25-diene-3B—O-glucoside, B-sitosterol glucoside, and a hypoglycemic substance charantin m.p. 266^0. A glycoalkaloid-vicine, was obtained from seeds.

60. *Morus alba* Linn. (Moraceae)

Common names
Eng.: White mulberry
Hindi: Tut
Beng.: Toot

Distribution

Cultivated throughout the plain parts of India.

Parts used: Leaves.

Pharmacological activities

Leaves produced antidiabetic activity diabetic animals. Marked hypoglycemic effect in normal and hyperglycemic subject was caused by the leaves extract.

Chemical constituents

See Antiferitility chapter.

61. *Mucuna prurita* Hook. (Leguminosae)

Common names
Eng.: Common Cowitch
Hindi: Kiwach
Beng.: Alkushi

Distribution

Found almost throughout India. The fine bristles on the pods cause very intense irritation, which sometimes last for several hours. Flowers are purple in auxiliary pendulous racemes.

Parts used: Seeds and fruits.

Pharmacological activities

Seeds produced hypoglycemic activity in normal albino rats. Fruits also caused hypoglycemic effect in albino rats.

Chemical constituents

Indole alkyamines.

62. *Murraya koenigii* spreng. (Rutaceae)

Common names
Eng.: Curry leaf tree
Hindi: Kurry patta, Gandhela, Mitha neem
Beng.: Kariaphulli.

Distribution

It is a small tree, found almost throughout India. It is cultivated also for its aromatic leaves used as flavoring agent of curries and chutney.

Parts used: Leaves.

Pharmacological activities

Aqueous extract of leaves, on oral administration showed hypoglycemic activity in normal and alloxan-diabetic dogs.

Chemical constituents

Leaves and fruits contain alkaloids koenimbine, m.p. 194^0, mahanimbine, m.p. 194^0 and koenigicine, m.p. 224^0. Moreover leaves contain mahanimbidine, cyclomaha nimbine, curryangine. Murrayzolidine, m.p. 168^0 mukoeic acid, m.p. 242^0 and girinimbine, m.p. 176^0 were obtained from bark.

63. *Musa sapientum* Linn. (Musaceae)

Common names
Eng.: Edible Banana
Hindi: Kela
Beng.: Kala

Distribution

It is native in India, and cultivated for its fruits.

Parts used: Flowers.

Pharmacological activities

Flowers produced hypoglycemic activity in rabbits. Maximum degree of hypoglycemic acvitity was caused in normal rabbits by the chloroform extractive fraction of flowers.

Significant hypoglycemic effect was produced both in normal as well as alloxan diabetic rats by the pectin, isolated from the juice of the inflorescence stalk of plantain.

Chemical constituents

See anti-ulcer chapter.

64. *Nigella sativa* Linn. (Ranunculaceae)

Common names
Eng.: Small Fennel
Hindi: Kalajira, Kalonji
Beng.: Kalajira.

Distribution

A small herb of 40-60 cm high, cultivated in Punjab, Bengal, Himachal Pradesh, Bihar and Assam. Seeds used as flavouring, carminative, and stimulant agents.

Parts used: Seeds.

Pharmacological activities

Volatile oil of seeds produced significant

hypoglycemic effects in fasting, normal and alloxan diabetic rabbits.

Chemical constituents

Seeds contain essential oil, Nigellone was isolated from essential oil. Seeds oil contains stigmasterol, cholesterol, α-spinasterol, campesterol and Beta-sitosterol.

65. *Nymhaea nouchali* Burm. F. (Nymphaeaceae)

Common names
Eng.: Indian Red Water-Lily
Hindi: Koka, kanval
Beng.: Shaluk

Distribution

An aquatic herb, common through the hotter parts of India. Flowering stalks used as vegetable, starchy rhizomes eaten raw or boiled.

Parts used: Roots.

Pharmacological activities

Rhizomes are used in dyspepsia, dysentery and diabetes. Roots exerted hypoglycemic activity in rabbits.

Chemical constituents

Root contains tannic, gallic acid, gum and starch.

66. *Ocimum sanctum* Linn. (Lamiaceae)

Common names
Eng.: Sacred Basil
Hindi: Tulsi
Beng.: Tulasi

Distribution

A strongly scented shrub. It is cultivated throughout India, and also considered a sacred plant to the Hindus. This plant is of two types, one is purple type, called Krishna tulsi, and another is green type, called Sri tulsi.

Parts used: Leaves.

Pharmacological activities

Leaves are used as stimulant, expectorant and used in bronchitis. The essential oil of leaves has antibacterial property.

Hypoglycemic activity was produced by the aqueous decoction of whole plant.

Chemical constituents

See Hepatoprotective chapters.

67. *Olea europaea* Linn. (Oleaceae)

Common names
Eng.: Common Olive.

Distribution

An evergreen tree, grown in Northern India and elsewhere. Both green and ripe fruits are edible. Mature fruits yield the fixed oil, known as Olive oil. The plant is native of Mediterranean region.

Parts used

Seeds and Leaves.

Pharmacological activities

Hypoglycemic effect was produced in normal human beings by Olive oil. Hydro-alcoholic extract of leaves lowered blood sugar level in rabbits.

Chemical constituents

Root bark contains, oleuropeic acid, the aglycone of 6-O-oleuropeoyl sucrose, characterized as {4-(1-hydroxyisopropyl)-1-cyclohexene-1-carboxylic acid}. A triterpenic acid, m.p. 240^0 was isolated from olive cakes. Fruits, when just ripe contain oil.

68. *Orchis mascula* Linn. (Orchidaceae)

Common names
Eng.: Salep Orchid
Hindi: Salap
Beng.: Salabmisri

Distribution

Tuberous roots, found at Western Himalayas. It is indigenous to Afghanistan and was imported to many places in India. Tuberous roots are used in medicine and it yields a large quality of mucilage with water and form a jelly.

Pharmacological activities

It is given in chronic diarrhea, dysentery and diabetes.

Chemical constituents

Tuber contains starch, mucilage, glycoside, trace volatile oil, sugar, a bitter substance and albumen.

69. *Orthosiphon spiralis* Merrill. (Lamiaceae)

Common names
Eng.: Kidney Tea Plant.
Distribution

A herb found at southern India, Madhya Pradesh, and Assam.

Pharmacological activities

Leaves extract reduced the blood sugar levels in diabetic patients but not consistently.

Chemical constituents

Leaves contain tannin, essential oil and potassium salts, 4',5,6,7-tetramethoxyflavone, 5-hydroxy-6,7,3',4'-tetramethoxyflavone and isosinensetin were isolated.

70. *Pinus roxburghii* Sarg. (Pinaceae)

Common names
Eng.: Chir Pine
Hindi: Chir
Sans.: Sarala

Distribution

An evergreen resinous tree, found at outer Himalayan ranges. Oleoresin, obtained from them, yields turpentine oil.

Parts used: Barks and roots.

Pharmacological activities

Bark and roots produced hypoglycemic activity in rabbits.

Chemical constituents

Longicyclene, dl-longifolene, Swedish turpentine a-longipinene and isopimaric acid were isolated. B-Sitosterol, friedelin and ceryl alcohol were isolated from bark.

71. *Portulaca oleracea* Linn. (Portulacaceae)

Common names
Eng.: Common Purslane
Hindi: Launinonia
Beng: Baraloniya.

Distribution

A herb, usually succulent, flowers are terminal, surrounded by a whole of leaves. It is found throughout India and used as a vegetable.

Parts used: Whole plant.

Pharmacological activities

Oral administration showed lowering of blood sugar in alloxan-diabetic rabbits.

Chemical constituents

4-(3,4-dihydroxyphenyl) alanine were detected.

72. *Prunus persica* Batsch. (Rosaceae)

Common names
Eng.: Peach
Hindi: Shaftalu, Aru.

Distribution

It is cultivated mainly in Northern India Fruits are edible.

Parts used: Leaves.

Pharmacological activities

Leaves produced hypoglycemic activity in rabbits and dogs.

Chemical Constituents

Aromadendrin, naringenin, 5,3′-dihydroxy-7.4′-dimethoxyflavone, m.p. 163^0 were isolated from bark. Leaves and flowers contain kaempferol and kaempferol-3=galactoside respectively. 3,4′5-trihydroxyflavanone-7-Beta-D-glucopy-ranoside, taxifolin-7-glucoside, 6-O-(Beta glucuronopyranosyl)-D-galactose, 2-O-(Beta glucuronopyranosyl)-D-mannose, galactose, xylose and rhamnose were detected in gum.

73. *Pterocarpus marspisum* Roxb. (Leguminosae)

Common names
Eng.: Indian Kino Tree
Hindi: Bija
Beng.: Piyasala, Pishal

Distribution

A tree of moderate to large size, up to 25 to 30 metre high. It is found mostly at Madhya Pradesh, Orissa, Bihar and Gujarat. The tree yields a gum-resin known as Kino.

Parts used: Bark and wood.

Pharmacological activities

Epicatechin, an active principle of the water extract of bark increased insulin release in rats.

Lowering of blood sugar was caused in dogs by the pterostilbene, constituents of wood.

Hypoglycemic effects were produced in rabbits by the alcoholic as well as aqueous extract of the heart wood.

Aqueous extract of wood resisted glucose absorption in mice. Aqueous extract of the drug produce hypoglycemic effect in rabbits.

No effect was exerted on the blood sugar of alloxan-rabbits and rats by the extract of wood.

Significant hypoglycemic effect with increase glucose tolerance was caused in rabbits by the alcoholic extract of the drug.

Lowering in the blood sugar levels in diabetic human beings as well as increase glucose tolerance, without any side effect, occurred by the decoction of the bark.

An aqueous infusion of wood increased glucose tolerance in rats.

Bark powder lowered the blood sugar levels in diabetes mellitus.

Hypoglycemic activity was produced in alloxan-diabetic rats by the decoction of the bark.

Induced hyperglycemia was prevented in rats by aqueous infusion of wood.

Reduction in urine sugar but without any affect on blood sugar levels of diabetic patients were caused by the aqueous extract of heart wood.

Chemical constituents

The main constituents are alkaloid and resin.

74. *Punica granatum* Linn. (Punicaceae)

Common names
Eng.: Pomegranate
Hindi: Anar
Beng.: Dalim

Distribution

Small tree, generally cultivated throughout India. Bark and fruit shells are used for tanning.

Parts used: Fruits rind.

Pharmacological activities

Final extract exhibited hypoglycemic effect in mild diabetic albino rats.

Chemical constituents

See antifertility chapter.

75. *Quercus infectoria* Olivier. (Fabaceae)

Common names
Eng.: Gall Oak
Hindi: Muphal, majuphal
Beng.: Majuphal.

Parts used: Galls

Pharmacological activities

Hypoglycemic action was produced in rabbits by a fraction of gall extract.

Chemical constituents

Ellagic acid, gallic acid, sitosterol, methyl oleanolic, methyl betulate and syringic acid were isolated from galls.

76. *Quercus ilicifolia* Roxb. (Fagaveae)

Common names
Assam: Bucklai
Nepal: Patlekatus

Distribution

It is found at Manipur, Khasi Hills, Sikkim, Bhutan and at the Himalayas at an altitude of 300-600 metre.

Parts used: Stem, bark

Pharmacological activities

A significant hypoglycemic activity was

exhibited in albino rats by the stem bark extract.

Chemical Constituents

Bark, leaves and wood contain tannin, Ferulic acid, canophyllal, canophyllol, friedelin maslinic acid, lignoceryl ferulate and lignoceryl alcohol were isolated.

77. *Rauvolfia serpentina* Benth. Ex. Kurz. (Apocynaceae)

Common names
Eng.: Serpentine wood
Hindi: Chandrabhaga
Beng.: Sarpagandha, Chandra.

Distribution

A small shrub found from the Himalayas southwards to Peninsular India.

Parts used: Roots

Pharmacological activities

Reserpine stimulated the hypoglycemic effect of insulin as well as the hyperglycemic effect of adrenalin also in normal subjects. Both the total extracts and reserpine suppressed the physiological hyperglycemic activities in diabetic patients.

Both the blood glucose and urine glucose were reduced in diabetic patients by the extracts. Moreover, the extracts also caused a sufficient reduction in blood sugar levels of cats.

Chemical constituents

See Antifertility chapater.

78. *Ricinus communis* Linn. (Euphorbiaceae)

Common names
Eng.: Castor seed, Castor oil plant
Hindi: Arandi
Beng.: Bherenda

Distribution

A small tree cultivated chiefly in Andhra Pradesh, Maharashtra, Karnataka and Orissa. It is also found as wild.

Parts used: Roots.

Pharmacological activities

Root extracts produced significant hypoglycemic activity in albino rats.

Chemical constituents

Stearic, palmitic, ricinoleic, arachidic, linoleic and oleic acid in castor oil as well as lupeol and 30-norlupan-3B-Ol-20-one in coat of castor bean were detected.

79. *Rivea cuneata* Wight. (Convolvulaceae)

Parts used: Leaves.

Pharmacological activities

Leaves are used in diabetes. Pulverized leaves produced a gradual fall in fasting blood sugar in rabbits, but oral administration showed little effect on fasting blood sugar in human beings. Significant hypoglycemic activity was caused by the alcoholic extracts of leaves in diabetic rats. Leaves did not produce any significant hypoglycemic activity in diabetic rabbits and rats.

Chemical constituents

Leaves contain glycoside.

80. ***Rourea santaloides*** **W. &A. (Connaraceae)**

Common names
Mah.: Wakeri
Bom.: Vardara
Beng.: Vidhadaki.

Parts used: Roots.

Pharmacological activities

Root is used as a bitter tonic in pulmonary complaints, scurvy, rheumatism and diabetes.

81. ***Salacia macrosperma*** **Wight. (Hippocrateaceae)**

Common names
Mar.: Lendphal
Mal.: Anakoranti

Distribution

Small tree, flowers are small, fruits are edible and sweet, found at Konkan and South wards in Western Ghats.

Parts used: Leaves and roots.

Pharmacological activities

Leaves and roots produced significant hypoglycemic effect in rabbits.

Chemical constituents

Salacia quinonemethide, tingenone, hydroxytingenone, pristimerin and three quinones saptarangi quinones A, B & C were isolated from root bark. A triterpene-salaspermic acid was also isolated.

82. ***Saussurea lappa*** **C.B. Clarke. (Compositae)**

Common names
Eng.: Kuth
Hindi: Kur, Kutha
Beng.: Kur

Distribution

A tall, perennial herb, found in Kashmir, and North-West Himalayas at an altitude about 2000 to 3000 meter. Flowers are purple.

Parts used: Roots.

Pharmacological activities

Alcoholic extract of roots showed a significant hypoglycemic response without any increase in plasma insulin in albino rats for a treatment of 7 days.

Chemical constituents

Leaves contain taraxasterol, m.p. 222^0 and taraxasterol acetate, m.p. 246^0, plant contains alkaloids, resins essential oil, 22, 23-dihydrostigmasterol and costol. Roots contain an essential oil, a glycoside and an alkaloid saussurine.

83. ***Scoparia dulcis*** **Linn. (Scrophulariaceae)**

Common names
Eng.: Sweet Broomweed
Santal: Yashtimadhu

Distribution

A herb, common in India on waste-land. Flowers are small and white, branches are twiggy.

Parts used: Stems and leaves.

Pharmacological activities

"Amellin", the active principle of leaves and stems decreased the high blood glucose levels, urine glucose and urine volume of diabetic patients to normal.

On application of fresh leaves with little cold water twice daily for four months to a patients, produced sugarless urine.

Partial reduction of urine glucose and urine volume were caused on rats, by the methanol and hexane extracts of leaves and stems.

Chemical constituents

Dulcitol, amellin from aerial parts, and mannitol and a triterpene, m.p. 288^0 from roots were isolated.

84. *Securigera securidaca* Linn. (Fabacaae)

Distribution

Native of some parts of Europe and Mediterranean region. In India, its cultivation at Bihar and West Bengal has been reported.

Parts used: Seeds.

Pharmacological activities

Hyperglycemia occurred significantly in mice after administration of seeds.

Aqueous extract of seeds exhibited hypoglycemic activity in anesthetized cats.

Chemical constituents

Cardenolides securidaside, m.p. 192^0 were isolated from seeds.

85. *Spathodea campanulata* Beauv. (Bignoniaceae)

Common Names
Eng.: African Tulip tree
Hindi: Rugtoora, Tel. Patode

Distribution

It is an evergreen tree, generally grown for its shade in many places, Flowers are large, bell-shaped and of orange scarlet color.

Parts used: Stem bark.

Pharmacological activities

Decoction of stem bark produced hypoglycemic activity in streptozocin diabetic mice.

Chemical constituents

Quercetin and caffeic acid were isolated from leaves.

86. *Strychnos potatorum* Linn. (Loganiaceae)

Common names
Eng.: Clearing-nut tree
Hindi: Neimal
Beng.: Nirmali

Distribution

This tree is found in Bengal, Continent and Southern India. Seeds are used for clearing muddy water, contain brucine but no strychnine, are non-poisonous.

Parts used: Seeds.

Pharmacological activities

Seeds are used in gonorrhea, diarrhea and diabetes.

Chemical Constituents

Leaves contain triterpene-isomotiol (fern-8-en-3Beta-ol), leaves and bark contain mixtures campesterol, stigmasterol and sitosterol. Alkaloid-diaboline, stigmasterol, oleanolic acid, its 3Beta-acetate, Beta-sitosterol and saponin containing oleanolic acid, galactose, mannose were isolated from seeds. Beside these 2-hydroxy-4-methoxybenzoic, sinapic, vanillic, chlorogenic and syringic acids, free sugars such as galactose, mannose and quercetin were isolated.

87. *Swertia chirayita* Roxb. (Gentianaceae)

Common names
Eng.: Chirayita
Hindi & Beng.: Chirayita

Distribution

Annual herb, found from temperate Himalayas to the Khasi Hills. Leaves are usually opposite, flowers are bluish to whitish. It is used as stomachic and laxative.

Pharmacological activities

Hexane extract showed a significant lowering of blood sugar in albino rats.

Chemical constituents

Roots and aerial parts contain 1-hydroxy-3,7,8-trimethoxyxanthone decussating, 1-hydroxy-3,5,8-trimethoxyxanthone,1,8-dihydroxy-3-methoxyxanthone,1,3,8-trihydroxy-5-methoxyxanthone,1,3,7,8-tetrahydroxyxanthone, 1,3,6,7-tetrahydroxyxanthone-c2-B-D-glucose (mangiterin and 1,3,5,8-tetrahydroxyxanthone. Beside these swerchirin, swertianin, swertinin, B-sitosterol friedelin and isobellidifolin were isolated.

88. *Syzgium cumini* Linn. (Myrtaceae)

Common names
Eng.: Black plum, Jambolan
Hindi: Jam, Jambu
Beng.: Kalajam, Jam

Distribution

A tree, found throughout India mostly cultivated for its edible fruits, seed coat within which two cotyledons are distinct, loosely adheres with pericarp.

Parts used: Seeds.

Pharmacological activities

Hypoglycemia was caused in the diabetic rats to normal levels by an active principle from powdered seeds.

Seed extract produced hypoglycemic response in albino rats.

Fall in fasting blood sugar in experimental animals was caused by the aqueous extract of seeds.

No remarkable difference in blood sugar levels was observed in rabbits produced by seeds. The seeds exhibited its utility in diabetic patients.

Active principle from seeds exerted strong hypoglycemia activity in mice.

Ethanolic extracts of bark showed hypoglycemic effect in hyperglycemic rabbits.

Fruits and seeds showed hypoglycemic response in rabbits.

Powder of fresh seeds lowered the blood sugar levels in diabetic animals on oral administration.

Different doses of alcohol and acetone extracts of seeds produced significant hypoglycemic activity in alloxan diabetes albino rats on long term feeding.

Seeds extract caused hypoglycemic response in male rabbits.

Lowering of the blood sugar in alloxan diabetic rats was produced by ethanolic extract of seeds.

Chemical constituents

Ellagic acid, isoquercetrin, quercetin, acetyl oleanolic acid, myricetin, kaempferol and two other triterpenoids were detected in flowers. Seeds contain ellagic, gallic acids, corilagin related ellagitannins, 3-galloylglucose, 1-galloylglucose, quercetin, 3,6-hexahydroxydiphenyl glucose and its isomer 4,6-hexahydroxydiphenyl glucose , beside these ferulic, gallic, caffeic, ellagic, 3,3'4'-tri-O-methylellagic, and 3,4'di-O-methylellagic acids were detected in seeds. Friedelin, friedelinol, sitosterol, and its glucoside, quercetin, betulinic acid, kaempferol and its 3-O-glucose and sucrose were isolated from bark. Hentriacontane, heptacosane, octacosanol, triacontane, crotegolic and betulinic acid from leaves and delphinidin-3-gentiobioside and malvidin-3-laminaribioside two anthocyanin from fruits were isolated.

89. *Tecoma stans* Linn. (Bignoniaceae)

Common names
Tel.: Pachagotla
Tam.: Sona-Patti

Distribution

An erect shrub, commonly found in gardens, bear showy flowers in terminal racemes.

Parts used: Stem.

Pharmacological activities

Aqueous and alcoholic extracts of stem reduced fasting blood sugar in rabbits and rats on oral administraiom.

Significant lowering of the blood sugar was caused in rats by the fresh aqueous extract of stem.

Glucose tolerance in rats was suitably influenced by the aqueous decoction and alcoholic extracts of stem. These extracts also caused increased glucose tolerance.

Chemical constituents

Leaves contain tecostanine, m.p. 85^0 and tecomanine (tecomine), beside these 4-noractinidine, N-normethyl-skytanthine, boschniakine, boschniakine and few other alkaloids were isolated.

90. *Tinospora crispa* Linn. (Menispermaceae)

Common names
Hindi: Gulancha
Beng.: Goloncha

Distribution

A climbing shrub, found throughout the tropical parts of India, Known by the same regional names as *Tinospora cordifolia* and used medicinally.

Pharmacological activities

Aqueous extract of stem showed hypoglycemic activity in diabetic rats.

Chemical constituents

Two unidentified alkaloids one is hydroxyl compound m.p. 95^0, y-sitosterol, another sterol, volatile oil, sodium, potassium, calcium, iron, copper and zinc were isolated from leaves.

91. *Trifolium alexandrinum* Linn. (Fabaceae)

Common names
Eng.: Berseem, Egyptian clover.

Distribution

A herb found as wild in temperate regions, sometimes cultivated for green manure.

Parts used: Seeds.

Pharmacological activities

Hypoglycemic activities were produced in different types of diabetic subjects comparable to tolbutamide by the powdered seeds. Hypoglycemic response also exerted in both normal and alloxan diabetic rabbits by infusion of seeds.

Chemical constituents

Seeds contain fatty acids like stearic, palmitic, linoleic, linolenic acids etc and xanthosine.

92. *Trigonella foenum* Graecum Linn. (Leguminosae)

Common names
Eng.: Fenugreek
Hindi: Muthi, methi
Beng: Methi, methuka

Distribution

A herb, cultivated in Northern India and some other places in India. Leaves are used as fodder and vegetable, seeds are used as spice and condiment. Flowers are whitish. It is native of Southern Europe.

Parts used: Seeds.

Pharmacological activities

Alkaloid trigonelline hydrochloride with feed extract produced a mild, transient hypoglycemia in alloxdiabetic rats. Remarkable hypoglycemic activities were caused by nicotinic acid and nicotinamide.

Hypoglycemic activity was exhibited in rabbits by the seed extract.

Chemical constituents

See Antifertility chapter.

93. *Urtica dioica* Linn. (Urticaceae)

Common names
Eng.: Stinging-Nettle
Hindi: Bichhu booti, Bichu

Distribution

Annual herbs with stinging hairs found at several parts of Himalayas.

Pharmacological activities

Extracts, after purified with picric acid, on oral and parenteral application, did not reduce the blood sugar in rabbits though the extracts were effective.

Chemical constituents

Leaves contain a neutral and a acidic carbohydrate protein polymer which contain serine-O-galactoside glycopeptide bond.

94. *Xanthium strumarium* Linn. (Compositae)

Common names
Eng.: Burweed

Hindi: Gokhru, Chota-gokhru, banokra
Beng.: Chota-dhatura

Distribution

Annual herbs, found throughout the warmer parts of India, as generally as weeds.

Parts used: Seeds.

Pharmacological activities

A white crystalline glycoside, patented in USA, obtained from seeds. Showed hypoglycemic effect in rabbits.

A crystalline substance from seeds produced hypoglycemic response in rats.

Chemical constituents

Xanthumin, m.p. 100^0 in aerial parts, strumaroside, m.p. 290^0 and stigmasterol in fruits, beside these B-sitosterol glucoside, xanthatin, m.p. 109^0, xanthinin, m.p. 110^0 isoxanthanol and flavonoid A m.p. 209^0 were isolated.

95. *Zea mays* Linn. (Gramineae)

Common names
Eng.: Corn, Maize
Hindi: Makka, makai, bhutta
Beng.: Bhutta

Distribution

A tall, stout annual, monoecious grasses, widely cultivated throughout in India for Maize grain or corn, extensively used as food.

Parts used: Styles.

Pharmacological activities

Hypoglycemic effect of well comparable with that of crystalline insulin was produced in fasting rabbits by the decoction of macerated styles. Hypoglycemic activity was also caused in rabbits by a fermented preparation of styles.

Chemical constituents

Three substituted cinnamoyl hydroxylcitric acid, characterized 2-O-trans-p-coumaroyl-(2s,3s)-hydroxy citric acid, 2-O-tranferuloyl-(2S,3S)-hydroxycitric acid, and 2-O-trans-caffeoyl-(2S,3S)-hydroxycitric acid were detected. Besides these di-O-(indole-3-acetyl)-myo-inositol and tri-O-(indole-3-acetyl)-myo-inositol from kernel and B-sitosterol from oil were isolated. 2-(2-hydroxy-7-methoxy-1.4-benzoxazin-3-one)-B-D-glucoside m.p. 228^0 from root and 3-galactoside-cyanidol coumarate in corn pigment in Peruvian variety were detecte. Along with a highly odorous compound – geosmin and a major constituents 2-heptanol sixty one volatile compound were identified.

96. *Zingiber officinale* Rosc. (Zingiberaceae)

Common names
Eng.: Ginger
Hindi: Adrak
Beng. Ada

Distribution

A perennial herb, cultivated in west Bengal, Andhra Pradesh, Uttar Pradesh, Kerala and many other places in India for the rhizomes, used as spice and in medicines as carminative.

Parts used: Rhizome.

Pharmacological activities

Fresh juice of rhizomes exerted significant hypoglycemic activity in diabetic rabbits and rats. This juice also caused lowering of blood glucose level in normal animals.

Chemical constituents

See Antiulcer chapter.

97. *Zizyphus jujuba* Mill. (Rhamnaceae)

Common names
Eng.: Indian Jujube
Hindi: Baer
Beng.: Kul, Boroi

Distribution

A tree, found throughout India, in dry deciduous forests. It is also cultivated for the edible fruits.

Parts used: Leaves.

Pharmacological activities

Significant fall in blood sugar level, same as tolbutamide was produced by the alkaloid of leaves.

Chemical constituents

Leaves contain alkaloid and rutin, and leucocyanidin in bark, and Oleanolic acid, betulinic acid and leucopelar gonidin in wood were detected. Fruits contain N-nomuciferine, asimilobine and stepharine.

7

MEDICINAL PLANTS FOR ANTIPYRETIC

INTRODUCTION

From an evolutionary perspective, fever as symptom is in picture since millions of years. In many situations, the elevation of body temperature increase chances for survival. The growth and virulence of several bacterial species are impaired at high temperatures. Temperatures in the febrile range appear to increase the phagocytic and bactericidal activity of neutrophils and the cytotoxic effects of lymphocytes.

However, fever involves considerable "costs" to the host in addition to discomfort. For each elevation of body temperature by 1°C, there is an increase in O_2 consumption of 13% and increase in caloric and fluid requirements.

Children are prone to develop seizures with fevers, particularly if they have a history of seizures. A single episode of a temperature of 37.8°C (100°F) in the first trimester of pregnancy doubles the risk of neural tube defects in the fetus.

Fever

Fever is an elevation of body temperature above the normal circadian range as the result of a change in the thermo regulatory center located in the anterior hypothalamus. A normal body temperature is ordinary maintained, despite environmental variations, through the ability of the thermo regulatory center to balance heat production by the tissues (notably, muscles and the liver) with heat dissipation. With fever, the balance is shifted to increase the core temperature. Hyperthermia is an elevation of body temperature above the hypothalamic set point due to insufficient heat dissipation (*e.g.*, in association with exercise, perspiration-inhibiting drugs, or a hot environment).

Where as the "normal" temperature in humans has been said to be 37°C (98.6°F) on the basis of Wunderlich's original observations more than 120 years ago, the overall mean oral temperature for healthy individuals aged 18 to 40 years is actually 36.8 ± 0.4°C (98.2 ± 0.7°F) with a nadir at 6 AM and a zenith at 4 to 6 PM. The maximum normal oral temperature at 6 AM is 37.2°C (98.9°F), and the maximum normal oral temperature at 4 PM is 37.7°C (99.9°F) both values defining the 99th percentile for healthy individuals. Given these criteria, an AM temperature of greater than 37.7°C (99.9°F) would define a fever. Rectal temperatures are generally 0.6°C (1°F) higher.

Causative Agent

Substances that cause fever are called pyrogens and may be either exogenous or endogenous. Exogenous pyrogens come from outside the host, whereas the host produces endogenous pyrogens, generally in response to initiating stimuli that are usually triggered by infection or inflammation. The majority of exogenous pyrogens are micro organisms, their products or toxins.

Endogenous pyrogens are polypeptides produced by a variety of host cells particularly monocytes/macrophages. Endogenous pyrogens produced either systemically or locally, gain entrance to the circulation or produce fever at the level of the thermoregulatory center of the hypothalamus. Interleukins and Cytokines are observed as the Endogenous pyrogens.

Mechanism of Temperature Control

(a) Body temperature is controlled by the hypothalamus. Neurons in both the preoptic anterior hypothalamus and the posterior hypothalamus receive two kinds of signals - one from peripheral nerves that reflect receptors for warmth and cold and the other from the temperature of the blood. These two signals are integrated by the thermo regulatory center of the hypothalamus to maintain normal temperature. In a neutral environment, the metabolic rate of humans consistently produces more heat than is necessary to maintain the core body temperature at 37°C. Therefore, the hypothalamus controls temperature by mechanisms of heat loss.

(b) Clusters of neurons in the preoptic/ anterior hypothalamus are supplied by a rich and permeable vascular network with limited blood brain barrier function. The specialized vascular network is called the organum. It is likely that the endothelial cells of this network release arachidonic acid metabolites when exposed to endogenous pyrogenic cytokines from the circulation. The arachidonic acid metabolites – mainly prostaglandin E_2 (PGE_2) - then presumably diffuse into the preoptic/anterior hypothalamic region and initiate fever.

(c) PGE_2 or other arachidonic acid products induce a second messenger such as cyclic AMP, which in turn raises the thermoregulatory set point. PGE_2 is believed to mediate the rise in the thermoregulatory set point. With the new, higher "thermostatic setting", signals go to various efferent nerves, particularly those sympathetic fibers innervating the peripheral blood vessels, which in turn initiate vasoconstriction and promote heat conservation. The thermo regulatory center also sends signals to the cerebral cortex, initiating behavioral changes such as seeking a warm environment, putting on more clothes, and special posturing.

(d) With the shunting of blood from the periphery and these behavioral changes, the body temperature usually rises by 2 to 3°C; if the hypothalamus calls for more heat, shivering (involuntary muscle contraction) is triggered to increase heat production. The combination of heat conservation and increased heat production continues until the temperature of the blood bathing the anterior hypothalamic neurons matches the new "setting." At that point, the hypothalamus maintains the new febrile temperature.

The hypothalamic set point is reset downward by the disappearance of stimulating pyrogenic cytokines or by the inhibition of local prostaglandin synthesis by cyclo-oxygenase inhibitors.

Factors Affecting Body Temperature

1. Diurnal variation

It is highest in the evening and lowest in the early hours of the morning (after the night rest). In the night workers, the rhythm is reversed. The average range is 1°F (0.55°C) to 1.5°F (0.83°C).

2. Age

In infants, regulation is imperfect effect. Hence, range of variation is wider. A fit of crying may raise and a cold bath may lower the body temperature. In old age, the body temperature may be subnormal due to low Basal Metabolism Rate.

3. Size

Heat production and heat loss depend upon the ratio of mass to body surface area. In a mouse, heat production is 452 large calories per kilogram body weight per 24 hrs whereas in a horse, it is only 14.5 large calories

4. Sex

In females, the body temperature may be a little lower. This is due to relatively low Basic Metabolism Rate and thick layer of subcutaneous fat. During menstruation, temperature slightly falls (0.3°F or 0.17°C) then it gradually rises and becomes maximum 24 to 48 hours after the ovulation. This rise is due to progesterone level of blood, which is secreted by the corpus luteum. Regular record of oral temperature in early morning is sometimes used to detect the exact date of ovulation in a woman, in clinical practice.

5. Food

Protein food, due to high SDA may raise body temperature. The act of ingestion of food may also raise body temperature.

6. Exercise

Increases temperature (Only 25% of muscular energy) is converted into mechanical work, the rest comes out as heat.

7. Atmospheric conditions

Temperature, humidity and movement of air are directly concerned with the amount of heat loss from the surface and thus affect body temperature.

8. Cold and warm baths

These have far greater influence than air at the same temperature but since duration of exposure to these baths are short; they have little effect on the normal body temperature. However, body temperature may remain elevated for a considerable time after a prolonged hot water bath.

9. Sleep

Because of muscular inactivity, sleep results in slight fall of body temperature.

10. Emotion

Body temperature may rise due to emotional disturbances. The rise of temperature may be as high as 2°C.

11. General anesthetics or Chlorpromazine

Tubocurarine paralysis the skeletal muscle, reduces heat production and results in fall temperature. Antipyretic drugs *e.g.*, sodium salicylate and aspirin antagonize the action of pyrogen on the hypothalamus, induce cutaneous vasodilatation and sweating and they are by help in reduction of temperature in febrile states.

Signs of Fever

Not surprisingly, many features associated with fever, including back pain, generalized myalgias, arthralgias, anorexia, and somnolence can be reproduced by infusions of purified cytokines. These symptoms may be reduced by cyclo-oxygenase inhibitors. Chills, a sensation of cold occurring in most fevers, are part of the central nervous system (CNS) response to the thermo regulatory set point's call for more heat.

Sweats occur with the activation of heat-loss mechanisms by antipyretic treatment, by attainment of a new "thermal ceiling chilliness, of the febrile stimulus. The intermittent administration of antipyretics may exaggerate swings of temperature, thereby causing chilliness, discomfort and exhaustion. Hypothalamic reflexes trigger sweating, which allows rapid dissipation of heat by evaporative loss.

Fever may be sustained, intermittent, remittent or relapsing. A sustained fever is one in which temperature elevation is persistent, with minimal variation. With intermittent fever, there is an exaggeration of the normal circadian rhythm; when this variation is extremely large, the fever is termed hectic or septic. Intermittent, hectic and septic patterns are common in deep-seated or systemic infections, malignancy and drug fevers. Remittent fever, in which the temperature falls each day but not to normal, is typical for tuberculosis, viral diseases, many bacterial infections and non infectious conditions.

Body temperature rises due to derangement of heat regulating mechanism. Toxins (pyrogens) act on WBC and produce endogenous pyrogen. This acts directly on anterior hypothalamus and the body temperature is elevated. Fever occurs due to any of the causes such as infections (*e.g.*, pneumonia, typhoid, fever etc), injury to nervous centers, dehydration, tissue destruction, administration of some drugs etc. In some fever *e.g.* malaria, shivering or rigor occurs.

1. Metabolism increases.
2. Blood pressure, pulse rate and cardiac output increases.
3. Rate of respiration increases.
4. There occurs negative nitrogen balance.

PLANTS FOR ANTIPYRETIC

1. *Aconitum ferox.* (Ranunculaceae)

Common names

Eng.: Bish; Indian aconite; Monkshood
Hindi & Beng.: Mithazahar; Bish and Katbish or Mitha Bish

Distribution

It is found in India and Nepal.

Parts used: Dried roots

Pharmacological activities

Antipyretic, antidiabetic, diaphoretic and diuretic.

Chemical constituents

Nepelline or pseudo-aconitine and aconitine.

2. *Alangium lamarckii* Thwaites. (Cornaceae)

Common names
Eng.: Sage-leaved alangium
Hindi & Beng.: Akola; Dhera and Akar-Kanta

Distribution

It is found in South India.

Parts used: Root, seed and leaves.

Pharmacological activities

Antipyretic, leprosy and antidote.

Chemical constituents

Bitter alkaloids, alangine.

3. *Alstonia scholaris* R. (Apocynaceae)

Common names
Eng.: Dita bark
Hindi & Beng.: Datyuni; Chhatiun and chhatim.

Distribution

It is found through out the in India

Parts used: Leaves, bark, milky Juice.

Pharmacological activities

Antipyretic, stimulant, carminative and aphrodisiac.

Chemical constituents

Bark contain alkaloid diamine and echitamine, echitenine also echicaoutchin and echitin.

4. *Anacardium occidentale* Linn. (Anacardiaceae)

Common names
Eng.: Cashew
Hindi & Beng.: Kaju and hijlibadam.

Distribution

It is found coast forests of India and all over South India.

Parts used: Fruit, seeds, spirit, bark and oil.

Pharmacological activities

Antipyretic, astringent and irritant.

Chemical constituents

Acrid oil (cardol), anacardic acid and bassorin.

5. *Andrographis paniculata* Nees. (Acanthaceae)

Common names
Eng.: Kiryat; Chiretta
Hindi & Beng.: Kiryat; Mahatita and Kalomegh.

Distribution

India; Bengal.

Parts used: Whole herb.

Pharmacological activities

Antipyretic and anthelmintic

Chemical constituents

Andrographolide and Kalmegh.

6. *Annona muricata* Linn. (Annonaceae)

Common names
Eng.: Sour sop of America.
Hindi & Beng.: Mamaphal.

Distribution

It is indigenous to West Indies but cultivated in the Bombay Presidency and Eastern India.

Parts used: Leaves, bark, root, seed, fruit.

Pharmacological activities

Antipyretic, purgative and astringent.

Chemical constituents

Root bark is given in ptomaine- poisoning especially after putrid fish eating.

7. *Emblica officinalis* Gaertn. (Euphorbiaceae)

Common names
Eng.: Amla; Indian gooseberry.
Hindi & Beng.: Amla; Amlika; Aonla; Usiriki and Amlaki.

Distribution

It is found all over India.

Parts used: Fruits, seeds, leaves, root, bark and flower.

Pharmacological activities

Antipyretic, antimutagenic, radioprotective, chemopreventive activity, hepatoprotective activity, cardioprotective activity and antidiabetic.

Chemical constituents

Gallic acid, tannins and resin.

8. *Azadirachta indica* Var. (Meliaceae)

Common names
Eng.: Neem, Margosa, Neem Tree.
Hindi & Beng.: Neem and Nim.

Distribution

It is native to India and the Indian subcontinent including Nepal, Pakistan, Bangladesh and Sri Lanka. It is typically grown in tropical and semi-tropical regions. Neem trees now also grow in islands located in the southern part of Iran.

Parts used: Leaves

Pharmacological activities

Antipyretic, antihelminthics, antifungal, antidiabetic, antibacterial, antiviral, contraceptive and sedative.

Chemical constituents

Azadirachtin, nimbidin and nimbinin.

9. *Berberis aristata* Dc. (Berberidaceae)

Common names
Eng.: Indian or Nepal or Ophthalric Barberry; Tree turmeric; False Calumba.
Hindi & Beng.: Rasaut; Chitra and Darhaldi.

Distribution

It is found in India and Bhutan.

Parts used: Root, bark, stem, wood

Pharmacological activities

Antipyretic, tonic, stomachic, antiperiodic and astringent. Root is purgative.

Chemical constituents

Root and wood are rich in a yellow alkaloids, berberine, bitter substance

10. *Cadreia toona* (Meliaceae)

Common names
Eng.: Toona

Distribution

It is found all over the India.

Parts used: Bark, gum, flowers.

Pharmacological activities

Antipyretic and astringent.

Chemical constituents

Coumarins, flavonoids, phytosterol, phenols, tenins, alkaloids, triterpenes and anthraquinones.

11. *Centella asiatica* Urban. (Umbelliferae)

Common names
Eng.: Brahmi, Asiatic Pennywort, Gotu Kola
Hindi & Beng.: Kula Kudi and Thol Kuri.

Distribution

It is found throughout the India

Parts used: Whole Plant, aerial parts, roots.

Pharmacological activities

Antioxidant and cytotoxic activities, antipyric and blood purifier. It has been used for treat in bronchitis, asthma, excessive secretion of gastric juices, dysentery, kidney trouble and dropsy in many communities.

Chemical constituents

Pentacyclic triterpenoids

12. *Cinchona officinalis* Hook. (Rubiaceae)

Common names
Eng.: Cinchona bark; Peruvian bark; Jesuit bark.

Distribution

It is found all over India.

Parts used: Bark

Pharmacological activities

Antipyretic.

Chemical constituents

Alkaloids, Viz quinine, cinchonine, quinidine, cinchonidine and hydroquinone. Barks also consist of chinic acid, chinovic acid and cincho-tannic acid.

13. *Cissampelospareira* Linn. (Menispermaceae)

Common names
Eng.: Velvet-leaf
Hindi & Beng.: Nirbisi, Harjori and Akanadi, Narbisi.

Distribution

Cultivated in tropical and sub tropical India from Sind and Punjab to South India and Ceylon.

Parts used: Roots, bark.

Pharmacological activities

Antipyretic, antilithic, diuretic, astringent and sedative.

Chemical constituents

Sepeerine, Bebeerines, Cissampeline or pelosine.

14. ***Cocculus cordifolius*** **Miers. (Menispermaceae)**

Common names
Eng.: Heart leaved moonseed
Hindi & Beng.: Gulancha

Distribution

Western India, Burma and Ceylon.

Parts used: Stem, leaves, roots

Pharmacological activities

Bitter tonic, stomachic, demulcent, antipyretic and aphrodisiac.

Chemical constituents

Root and stem contain starch extract, bitter principle and a trace of berberine. Leaves are highly mucilage.

15. ***Coscinium fene stratum*** **Gaertn & Colebr. (Menispermaceae)**

Common names
Eng.: Tree turmeric.
Hindi & Beng.: Jhar Haldi and Haldi-gach.

Distribution

It is found all over the India.

Parts used: Stems

Pharmacological activities

Antipyretic and stomachic.

Chemical constituents

Saponin and berberine.

16. ***Cuscuta reflexa*** **Roxb. (Convolvulaceae)**

Common names
Eng.: Dodder
Hindi & Beng.: Akasbel and Algusi.

Distribution

It is found throughout the India.

Parts used: Seeds, stem, fruits.

Pharmacological activities

Antipyretic and carminative.

Chemical constituents

Quercetin, resin and and cuscutine.

17. ***Grewia asiatica*** **Linn. (Tiliaceae)**

Common names
Eng.: Parushaka, Parusha, Parapara, Alpasthi, Dhanvanachchhada.
Hindi & Beng.: Dhanvan anachada and Phalsa.

Distribution

It is found throughout greater part of India, in salt range of Punjab, Western Himalaya up to 1000 m, N. Bengal, Bihar, Chota Nagpur, Orissa, Gujarat, Konkan, Deccan and South India.

Parts used: Bark, Leaves

Pharmacological activities

Antipyretic, rheumatism, demulcent. Spasmolytic, hypotensive, antidiabetic, antifertility and antibacterial.

Chemical constituents

Triterpenoid-erythrodiol, taraxasterol, p-sitosterol, lupeol, betulin, lupenone, friedelin, P-amyrin (stem bark); p-sitosterol (heartwood) glutaric acid, phenylalanine, glucose, xylose, arabinose, tannins- catechin and leuco anthocyanins (fruit); palmitic, stearic, oleic, linoleic acid and sitosterol (seeds); quercetin, kaempferol and mixture of their glycosides (leaves).

18. *Hemidesmus indicus* Linn. (Asclepiadaceae)

Common names
Eng.: Indian sarsaparilla
Hindi & Beng.: Sungandhi

Distribution

It is found throughout the in India.

Parts used: Root, Juice

Pharmacological activities

Antipyretic, diuretic, diaphoretic, tonic and demulcent.

Chemical constituents

The roots of *H. indicus* contain hexa-triacontane, lupeol, its octacosanoate, α-amyrin, β-amyrin, its acetate and sitosterol. It also contains new coumarinolignoid-hemidesminine, hemidesmin I and hemidesmin II50, six pentacyclic triterpenes including two oleanenes, and three ursenes. The leaves contain tannins, flavonoids, hyperoside, rutin and coumarino. Leuco derma lignoids such as hemidesminine, hemidesmin I and hemidesmin II are rare group of naturally occurring compounds present in leaves.

19. *Eclipta erecta* Linn. (Compositae)

Common names
Eng.: Bhringaraj
Hindi & Beng.: Bhangro Moch karnd and Kesooria.

Distribution

It is found throughout the India.

Parts used: Roots, leaves

Pharmacological activities

Antipyretic, purgative and emetic.

Chemical constituents

Resin and an alkaloidal principle ecliptine.

20. *Asparagus adscendens* Roxb. (Liliaceae)

Common names
Eng.: Shatavari
Hindi & Beng.: Safed musli

Distribution

It is found in India. Cultivated in Himalaya, Punjab from Murree to Kumaon, Bombay, Rohilkhand, Oudh and Central India.

Parts used: Tuberous roots, rhizome.

Pharmacological Activities

Antipyretic, demulcent, nutritive tonic, galactegogue and useful in the treatment of diarrhea, dysentery.

Chemical constituents

Saponins, Asparagine, albuminoids matter, mucilage and cellulose.

21. *Daemia extensa* R. Br. (Asclepiaaeae)

Common names
Eng.: Phala-Kantak
Hindi & Beng.: Utranajutuka; Utran and Chhagalbati; Chagulbanti

Distribution

It is found all over India.

Parts used: Leaves, roots and root bark.

Pharmacological activities

Antipyretic, expectorant and antihelmintic

Chemical constituents

Alkaloids named Daemine.

22. *Desmodium gangentium* DC. (Leguminosae)

Common names
Eng.: Dasa mularishtam
Hindi & Beng.: Sarivan

Distribution

Indian Himalayas.

Parts used: Root, Bark

Pharmacological activities

Roots are alterative, tonic, anthelmintic, aphrodisiac and astringent to the bowels; used in typhoid, fever, piles, asthma, bronchitis, dysentery, diarrhoea, biliousness and cough; also useful in chronic affections of the chest and lungs. Root extract is used in whooping cough. Alkaloids of the aerial parts are smooth muscle and CNS stimulant, and cardiac depressant, possesses anti cholinesterase activity.

Chemical constituents

Aerial parts contain five tryptamine derivatives, Nb-Me-tetrahydroharman and 6-OMe-2-Me- β-carboliniumcation. Root contains alkaloids, N, N-dimethyl-tryptamine and its N-oxide, N-methyltyramine, hypaphorine, candicine, hordenine and a β-phenylethylamine base. Roots also contain several carboxylated and decarboxylated tryptamine and β-alkylamine 1+derivatives, a pterocarpan, ptero-carpanoid gangetin, gangetinin and desmodin. Seeds contain β-carboline alkaloid, indole-3-alkylamine, carbolines and five phospholipids. Seeds showed presence of sugars, fatty oil and alkaloids.

23. *Hemidesmus indicus* R. Br. (Asdepiadaceae)

Common names
Eng.: Suganhi; Indian Sarsaparilla and
Hindi & Beng.: Magrabu; Salsa; Kalisar and Anantamul.

Distribution

It is found throughout the India.

Parts used: Roots, root- bark and juice.

Pharmacological activities

Antipyretic and demulcent.

Chemical constituents

Coumarin and hemidesmine.

24. *Lantana involucrat* Linn. (Verbenaceae)

Common names
Eng.: Lantana odorata.
Hindi & Beng.: White lantana, Wild Sage, Button sage.

Distribution

It is found in India.

Parts used: Fruit, seed, bark, oil

Pharmacological activities

Antipyretic, cytotoxic activities against various human tumor cell lines and irritant.

Chemical constituents

Isopropenylfurano-β-naphthoquinones, designated lantalucratins A, B and C.

Chemical constituents

25. *Tinospora cordifolia* Miers. (Menispermaceae)

Common names
Eng.: Cordifolium
Hindi & Beng.: Gurach and Gulancha

Distribution

South India.

Parts used: Stem, root

Pharmacological activities

Antipyretic, diuretic and antidote.

Chemical constituents

Bitter substance and berberine.

26. *Ocimum sanctum* Linn. (Labiatae)

Common names
Eng.: Tulsi, Sacred Basi, Holy Basil
Hindi & Beng.: Tulsi

Distribution

It is found all over India, Andaman and Nicobar Islands.

Parts used: Leaves

Pharmacological activities

Antipyretic, antitussive, culinary purposes, drugs flavouring, insecticide and Perfumery.

Chemical constituents

The plant contains mainly phenols, aldehydes, tannins, saponin and fats. The essential oil components are eugenol (about 71%, eugenol methyl ether (20%), nerol caryophyllene, selinene, α-pinene, β-pinene, camphor cineole, linalool and carvacrol (3%). A terpeneurobsolic acid possessing anticancer properties has also been isolated.

27. *Oldenlandia herbacea* Linn. (Rubiaceae)

Common names
Eng.: Indian Madder; "tan ts'ao
Hindi & Beng.: Daman-papper and Kitpapra.

Distribution

It is found all over India.

Parts used: Whole herb

Pharmacological activities

Antipyretic, irritability and nervous depression and also in chronic malaria as a good febrifuge.

Chemical constituents

Flavonoid, iridoids, alkaloids and triterpenes.

28. *Pavonia odorata* Willd. (Malvaceae)

Common names
Eng.: Harivera
Hindi & Beng.: Bala; sugandha and Bala; Bola.

Distribution

Western India.

Parts used: Roots

Pharmacological activities

Antipyretic and diuretic.

Chemical constituents

Palmitic acid, hexahydrofarnesyl acetone, β-eudesmol and β-caryophyllene oxide. The most characteristic aroma compounds in the volatile oil were identified for β-caryophyllene oxide.

29. *Peganum harmala* Linn. (Rutaceae)

Common names
Eng.: Syrian Rue
Hindi & Beng.: Hurmal; Harmal and Isband

Distribution

North India

Parts used: Seeds

Pharmacological activities

Antipyretic, abortifacient and stimulant.

Chemical constituents

It contain three alkaloids, harmalol, harmaline and harmine.

30. *Picorrhiza kurroa* Royle. (Scrophulariaceae)

Common names
Hindi & Beng.: Katuka

Distribution:

Picrorhiza kurroa is a small perennial herb that grows in hilly parts of India particular in Himalayas between 3000 and 5000 metres.

Parts used: Dried rhizome

Pharmacological activities

Antipyretic, asthma, liver damage, wound healing, vitiligo and laxative.

Chemical constituents

Chemical composition of Picrorhiza kurroa include Kutkin, a bitter glycoside which contains two C-9 iridoid glycosides-Picroside I and Kutakoside.

31. *Piper betle*. (Piperaceae)

Common names
Eng.: Betel leaf; Pepper
Hindi & Beng.: Pan

Distribution

It is found throughout the India.

Parts used: Leaves

Pharmacological activities

Carminative and antipyretic.

Chemical constituents

Volatile oil, chavicol, phenol and alkaloid.

32. *Piper nigrum* Linn. (Piperaceae)

Common names
Eng.: Black pepper
Hindi & Beng.: Kali-Mirch; Gulmirch and Kalimirch.

Distribution

Western India.

Parts used: Dried fruits

Pharmacological activities

Antipyretic, antifungal activity, Immunomodulatory activity, anti-cancer activity, carminative and antiperiodic.

Chemical constituents

Piperine, alkamides, piptigrine, wisanine, dipiperamide D, and dipiperamide E.

33. *Prunu spadus*. (Rosaceae)

Common names
Eng.: Birdcherry.
Hindi & Beng.: Jamana

Distribution

India; Bhutan.

Parts used: Seed; Oil.

Pharmacological activities

Antipyretic and diuretic.

34. *Pterocarpus santalinus* Linn. (Papilionaceae)

Common names
Eng.: Red sanders or Red sandalwood
Hindi & Beng.: Rakta-chandana

Distribution

It is found in South India.

Parts used: Wood

Pharmacological activities

Antipyretic, cooling, tonic and astringent.

Chemical constituents

Santalin or santalic acid, santalpterocarpin and homopterocarpin.

35. *Rubia cordifolia* Linn. (Rubiaceae)

Common names
Eng.: Indian Madder; Dyer's Madder
Hindi & Beng.: Manjith, Chitravalli

Distribution

North India.

Parts used: Roots

Pharmacological activities

Antipyretic, anti-acne property, anti-cancer property, Wound healing activity, anti-diabetic, astringent and diuretic.

Chemical constituents

Anthraquinone, terpenes, glycosides etc..

36. *Salvadora persica* Linn. (Salvadoraceae)

Common names
Eng.: Tooth brush tree
Hindi & Beng.: ChhotaPilu

Distribution

It is found throughout the India

Parts used: Root-Bark

Pharmacological activities

Antipyretic, carminative and purgative.

Chemical constituents

Salvadorine and trimethylamine.

37. *Santalum album* Linn. (Santalaceae)

Common names
Eng.: White sandalwood tree
Hindi & Beng.: Safed Chandan and Chandan; sadachandan.

Distribution

It is found in South India.

Parts used: Wood, volatile oil

Pharmacological activities

Antipyretic, bitter, cooling, genitourinary and bronchial tracts, sedative and astringent.

Chemical constituents

Resin, essential oil, santalol, tannin, santalol A and santalol B.

38. *Swertia chirata* Ham. (Gentianaceae)

Common names
Eng.: Chirata, Chiretta
Hindi & Beng.: Kiryat- Charayata; Jwaran-Thakah and Mahatita; Chireta.

Distribution

North India.

Parts used: Whole herb

Pharmacological activities

Hepatoprotective, antihepatotoxic, anti-microbial, anti-inflammatory, anti-carcinogenic, antileprosy, hypoglycemic, antimalarial, antioxidant, anticholinergic, CNS depressant and mutagenicity, antipyretic and antidote.

Chemical constituents

The major bioactives of Swertia are xanthones, however, other secondary metabolites such as flavonoids, iridoid glycosides and triterpenoids.

39. *Tamarindus indica* Linn. (Caesalpiniaceae)

Common names
Eng.: Tamarind tree
Hindi & Beng.: Imli; Amli and Tentul; Ambli; Tintil.

Distribution

It is found in South India.

Parts used: Fruits

Pharmacological activities

Antipyretic, laxative, digestive, antiscorbutic, antibilious and carminative.

Chemical constituents

Tartaric acid, citric acid, malic acid and acetic acid, tartaric of potassium, gum and pectin.

40. *Tecoma stans*. (Bignoniaceae)

Common names
Hindi & Beng.: Yellow Cedar

Distribution

India.

Parts used: Wood, oil

Pharmacological activities

Sedative and antipyretic.

Chemical constituents

Alklaloids, saponins and tannins.

41. *Terminalia bellirica* (Gaertn.) Roxb. (Combretaceae)

Common names
Eng.: Beleric, Belleric myrobalan
Hindi & Beng.: Bahera; Bahira and Bohera; Boheri; Buhuru.

Distribution

It is found in India.

Parts used: Fruit

Pharmacological activities

It is Astringent, Tonic, Expectorant and Laxative. It is used in coughs and sore throat. Its pulp is used in dropsy, piles and diarrhea. It is also useful in leprosy, fever and hair care. It is also used in oxalic acid and preparation of ink.

Chemical constituents

Ellagic acid, gallo-tannic acid and resin.

42. *Terminalia chebula* Retz. (Combretaceae)

Common names
Eng.: Myrobalan, Chebulic myrobalan; Ink-nut; Indian gall-nut.
Hindi & Beng.: Harar; Hardad and Hora; Haritaki.

Distribution

It is found throughout in India

Parts used: Fruit, roots, bark.

Pharmacological activities

It is astringent, antipyretic, purgative, stomachic and laxative. It is useful in asthma, piles and cough. It is also useful in healing of wounds and scalds. It is used as gargle against inflammation of mucous membrane of mouth.

Chemical constituents

Tannins, polyphenols, terpenes, anthocyanins, flavonoids, alkaloids and glycosides. It also consists of chebulagic acid, punicalagin, chebulanin, corilagin, neochebulinic acid, ellagic acid, chebulinic acid.

43. *Nyctanthes arbor-tristis* Linn. (Oleaceae)

Common names
Eng.: Night-flowering Jasmine.
Hindi & Beng.: Harshingar; Parijata and Shephali; Seuli.

Distribution

Nyctanthes arbor-tristis is widely distributed along subtropical, tropical to sub Himalayan regions in the South East Asia.

Parts used: Leaves

Pharmacological activities

Sciatica, arthritis, malaria, antidote, laxative, hepatoprotective, antileishmaniasis, antiviral, antifungal, antipyretic, antihistaminic, antimalarial, antibacterial, anti-inflammatory, antioxidant activities.

Chemical constituents

Tannins, flavonoids, essential oils, flavanol glycoside, oleanic acid, essential oils, tannic acid, carotene, friedeline, lupeol, glucose, benzoic acid.

44. *Trichosanthes dioica* Roxb. (Cucurbitaceae)

Common names
Eng.: Wild Snakegourd
Hindi & Beng.: Palwal; Parvar and Patol; Potal; Potol.

Distribution

North India.

Parts used: Fruits

Pharmacological activities

Antipyretic, purgative, cardiotonic, antiulcer and laxative.

Chemical constituents

The various chemical constituents present in *T. dioica* are vitamin A, vitamin C, tannins, saponins, alkaloids, mixture of novel peptides, proteins tetra and pentacyclic triterpenes, etc.

45. *Viola odorata* Linn. (Violaceae)

Common names
Eng.: Wild violet.
Hindi & Beng.: Banaphsa; Bag-banosa and Banosa.

Distribution

It is found all over India.

Parts used: Whole Herb

Pharmacological activities

Antipyretic, astringent, demulcent, aperient and antitussive.

Chemical Constituents

Violine, volatile oil, methyl salicylic ester, quercetin and glucoside.

46. *Vitex negundo* Linn. (Verbenaceae)

Common names

Eng.: Five–leaved chaste tree
Hindi & Beng.: Nirgandi; Sambhalu; Nisinda and Nishinda.

Distribution

It is found in South India and Burma.

Parts used: Roots, flower, fruits and bark.

Pharmacological activities

Antipyretic, expectorant, aromatic, bitter and astringent.

Chemical constituents

Essential oil, resin, fruit contain an acid resin, as astringent organic acid, malic acid, trace of an alkaloids and a coloring matter.

8

MEDICINAL PLANTS FOR ANTICANCER

INTRODUCTION

Cancer is a major public health burden in both developed and developing countries. It is an abnormal growth of cells in body that can lead to death. Cancer cells usually invade and destroy normal cells. These cells are born due to imbalance in the body and by correcting this imbalance, the cancer may be treated. Billions of dollars have been spent on cancer research and yet we do not understand exactly what cancer is. Every year, millions of people are diagnosed with cancer, leading to death. According to the American Cancer Society, deaths arising from cancer constitute 2 – 3% of the annual deaths recorded worldwide. Thus cancer kills about 3500 million people annually all over the world. Several chemo preventive agents are used to treat cancer, but they cause toxicity that restricts their usage. There are several medicines available in the market to treat the various types of cancer but no drug is found to be fully effective and safe. The major problem in the cancer chemotherapy is the toxicity of the established drugs. However plants and plant derived products have proved effective and safe in the treatment and management of cancers. These days most of the research work on cancer drugs is targeted on plants and plants derived natural products. Many natural products and their analogues have been identified as potent anti-cancer agents and day by day the anti -cancer property of various plants is being identified.

What is Cancer

Cancer is a general term applied of series of malignant diseases that may affect different parts of body. These diseases are characterized by a rapid and uncontrolled formation of abnormal cells, which may mass together to form a growth or tumor, or proliferate throughout the body, initiating abnormal growth at other sites. If the process is not arrested, it may progress until it causes the death of the organism. The main forms of treatment for cancer in humans are surgery, radiation and drugs (cancer chemotherapeutic agents). Cancer chemotherapeutic agents can often provide temporary relief of symptoms, prolongation of life, and occasionally cures. In recent years, a lot of effort has been applied to the synthesis of potential anticancer drugs. Many hundreds of chemical variants of known class of cancer chemotherapeutic agents have been synthesized but have a more side effects. A successful anticancer drug should kill or incapacitate cancer cells without causing excessive damage to normal cells. This ideal is difficult, or perhaps impossible, to attain and is why cancer patients frequently suffer unpleasant side effects when under-going treatment. Synthesis of modifications of known drug continues as an important aspect of research. However, a waste amount of synthetic work has given relatively small improvements over the prototype drugs. There is a continued need for new prototype-new templates to use in the design of potential chemotherapeutic agents: natural products are providing such templates. Recent studies of tumor-inhibiting compound of plant origin have yielded an impressive array of novel structures. Many of these structures are extremely complex, and it is most unlikely that such compounds would have been synthesized in empirical approaches to new drugs

Causes of Cancer

Modern medicine attributes most cases of cancer to changes in DNA that reduce or eliminate the normal controls over cellular growth, maturation, and programmed cell death. These changes are more likely to occur in people with certain genetic backgrounds (as illustrated by the finding of genes associated with some cases of cancer and familial prevalence of certain cancers) and in persons infected by chronic viruses (*e.g.*, viral hepatitis may lead to liver cancer; HIV may lead to lymphoma). The ultimate cause, regardless of genetic propensity or viruses that may influence the risk of the cancer, is often exposure to carcinogenic chemicals (including those found in nature) and/or to radiation (including natural cosmic and earthly radiation), coupled with a failure of the immune system to eliminate the cancer cells at an early stage in their multiplication. The immunological weakness might arise years after the exposure to chemicals or radiation. Other factors such as tobacco smoking, alcohol consumption, excess use of caffeine and other drugs, sunshine, infections from such oncogenic virus like cervical papilloma viruses, adeno viruses Kaposi's sarcoma (HSV) or exposure to asbestos. These obviously are implicated as causal agents of mammalian cancers. However a large population of people is often exposed to these agents. Consequently cancer cells continue to divide even in situations in which normal cells will usually wait for a special chemical transduction signal. The tumor cells would ignore such stop signals that are sent out by adjacent tissues. A Cancer cell also has the character of immortality even *in vitro* whereas normal cells stop dividing after 50-70 generations and undergoes a programmed cell death (Apoptosis). Cancer cells continue to grow invading near by tissues and metastasizing to distant parts of the body. Metastasis is the most lethal aspect of carcinogenesis.

Types of Cancers

1. Cancers of blood and lymphatic systems

(a) Hodgkin's disease
(b) Leukemias
(c) Lymphomas
(d) Multiple myeloma,
(e) Waldenstrom's disease

2. Skin cancers

(a) Malignant Melanoma

3. Cancers of digestive systems

(a) Esophageal cancer
(b) Stomach cancer
(c) Cancer of pancreas
(d) Liver cancer
(e) Colon and rectal cancer
(f) Anal cancer

4. Cancers of urinary system

(a) Kidney cancer

(b) Bladder cancer
(c) Testis cancer
(d) Prostate cancer

5. Cancers in women

(a) Breast cancer
(b) Ovarian cancer
(c) Gynecological cancer
(d) Choriocarcinoma

6. Miscellaneous cancers

(a) Brain cancer
(b) Bone cancer
(c) Carcinoid cancer
(d) Nasopharyngeal cancer
(e) Retroperitoneal sarcomas
(f) Soft tissue cancer,
(g) Thyroid cancer

The Mechanism on Cancer Therapy

1. Inhibiting cancer cell proliferation directly by stimulating macrophage phagocytosis, enhancing natural killer cell activity.
2. Promoting apoptosis of cancer cells by increasing production of interferon, interleukin-2 immunoglobulin and complement in blood serum.
3. Enforcing the necrosis of tumor and inhibiting its translocation and spread by blocking the blood source of tumor tissue.
4. Enhancing the number of leukocytes and platelets by stimulating the hemopoietic function.
5. Promoting the reverse transformation from tumor cells into normal cells.
6. Promoting metabolism and preventing carcinogenesis of normal cells.
7. Stimulating appetite, improving quality of sleep, relieving pain, thus benefiting patients health.

Environmental Factors

Environmental factors which, from a scientist's standpoint, include smoking, diet, and infectious diseases as well as chemicals and radiation in our homes and work place along with trace levels of pollutants in food, drinking water and in air. Other factors which are more likely to affect are tobacco use, unhealthy diet, not enough physical activity, however the degree of risk from pollutants depends on the concentration, intensity and exposure. The cancer risk becomes highly increase d where workers are exposed to ionizing radiation, carcinomas chemicals, certain metals and some other specific substances even exposed at low levels. Passive tobacco smoke manifold increase the risk in a large population who do not smoke but exposed to exhaled smoke of smokers.

Ayurvedic Concept of Cancer

Charaka and Sushruta Samhita both described the equivalent of cancer as "*granthi*" and "*arbuda*". "*Granthi*" and "*Arbuda*" can be inflammatory or devoid of inflammation, based on the doshas involved. Three doshas "Vata, Pitta and Kapha" in body are responsible for disease and the balanced coordination of these doshas in body, mind and consciousness is the Ayurvedic definition of health. Tridoshic arbudas are usually malignant because all three major body humors lose mutual coordination, resulting in a morbid condition. Neoplasm can be classified in Ayurveda depends upon various clinical symptoms in relation to tridoshas.

Group I

Diseases that can be named as clear malignancies, including *arbuda* and *granthi,* such as *mamsarbuda* (sarcomas) and *raktarbuda* (leukaemia), *mukharbuda* (oral cancer) and *asadhyavrana* (incurable or malignant ulcers).

Group II

Diseases that are not cancers but can be considered probable malignancies, such as ulcers and growths. Examples of these are *mamsaja oshtharoga* (growth of lips), *asadhya galganda* (incurable thyroid tumour*), tridosaja gulmas* and *asadhya udara roga* (abdominal tumours like carcinomas of the stomach and liver or lymphomas).

Group III

Diseases in which there is a possibility of malignancy, such as *visarpa, asadhya kamala* (incurable jaundice), *asadhya pradara* (in treatable sinusitis).

Role of Plants as Medicinal and Anti-cancer Agents

Plants, since ancient time, are using for health benefits by all cultures as well as source of medicines. It has been estimated that about 80–85% of global population rely on traditional medicines for their primarily health care needs and it is assumed that a major part of traditional therapy involves the use of plant extracts or their active principles. Although a lot of recent investigations have been carried out for advancements in the treatment and control of cancer progression, significant work and room for improvement remain. The main disadvantages of synthetic drugs are the associated side effects. However natural therapies, such as the use of the plants or plant derived natural products are being beneficial to combat cancer. The search for anti -cancer agents from plant sources started in the 1950s when discovery and development of the vinca alkaloids (vinblastin and vincristine), and the isolation of the cytotoxic podophyllotoxins was carried out.

PLANTS FOR ANTICANCER

1. *Allium sativum* (Liliaceae)

Common names
Eng.: Garlic
Hindi & Beng.: Lahsan, Lehsun

Distribution

It is found throughout in India - Cultivated mainly in Karnataka, Tamil Nadu and Andhra Pradesh and Gujarat.

Parts used: Bulbs

Pharmacological activities

Garlic is commonly used as a stimulant, diaphoretic, expectorant, diuretic, tonic, antioxidant properties and Garlic extract is found to inhibit growth of many cancers including those of the breast, bladder, skin, colon, oesophagus, stomach and the lung.

Chemical constituents

Alliin, allicinalliin, alliinase, S-allyl-cysteine (SAC), diallyl disulphide (DADS), diallyl-trisulphide (DATS) and methylallyl-trisulphide.

2. *Actinidia chinensis* Planch. (Actinidiaceae)

Common names
Eng.: China Goose berry, Kiwi fruit

Distribution

Kiwifruit or Chinese goose berry is rapidly become a major horticultural crop in New

Zealand, and plantings are increasing in many other countries.

Parts used: Root

Pharmacological activities

Root is used as immune-enhancing and cancer activity.

Chemical constituents

Vitamin C, Polysaccharide known as "ACPS-R"

3. *Aloe barbadensis* (Liliaceae)

Common names
Eng.: Aloe vera
Hindi & Beng.: Ghrita kumari and korpad

Distribution

It has been widely cultivated throughout the world. Grows mainly in the dry regions of Africa, Asia, Europe and America.

Parts used: Leaves

Pharmacological Activities

It is used as wounds, burns, various skin disease, laxative, constipation, ulcers, diabetes, headaches, arthritis, and coughs. Taking aloe internally does have side effects, It has been determined that it can also help with treating minor vaginal irritations.

Chemical constituents

Aloe-emodin, emodin, aloeacemaman, Barbaloin and Iso-barbaloin.

4. *Ananas comosus* (Bromeliaceae)

Common names
Eng.: Pineapple, Ananas
Hindi & Beng.: Ananas and Anaras

Distribution

It is cultivated throughout in India, and common in bazaar.

Parts used: Fruit and leaves

Pharmacological activities

Fresh juice of leaves is powerfully purgative, anthelmintic and vermicide. Juice of ripe fruit is antiscorbutic, diuretic, diaphoretic, aperients and refrigerant, fever, jaundice and anticancer activity.

Chemical Zonstituents

Bromelain, Vitamin A, B1, B6 and C.

5. *Angelica sinensis* (Umbelliferae or Apiaceae)

Common names
Eng.: Angelica, *dong quai* or "female ginseng"

Distribution

Indigenous to China. *Angelica sinensis* grows in cool high altitude mountains in China, Japan, and Korea. The yellowish brown root of the plant is harvested in fall and is a well-known Chinese medicine used over thousands of years.

Parts used: Root

Pharmacological Activities

It is used for women's health, cardiovascular condition, inflammation, headache, infections, osteoarthrosis, mild anemia, fatigue and high blood pressure, anticancer and has *in vitro* antioxidant activity.

Chemical constituents

Polysaccharide fraction of known as "AR-4" also consist phytosterol, ligustilit, b-butylphtalit, cnidilit, isoenidilit, π- Cymen, ferulate and flavonoids.

6. *Annona squamosa* (Annonaceae)

Common names:
Eng.: Custard apple, sweet Sop of America, Sugar apple
Hindi & Beng.: Sharifah, Sitaphal and Ata

Distribution

In gardens all over India

Parts used: Leaves, bark, root, seed and fruit

Pharmacological Activities

Bark is powerful, astringent and tonic. Leaves and seeds and unripe fruit are vermicide or insecticide, diarrhea and dysentery. Leaves are anthelmintic. Root is a violent purgative and seed are detergent. Ripe fruits are used for Ulcer and applied to malignant tumors. Leaves are tumors, cancers root.

Chemical Constituents

Acetogenins and fruit contain acrid principle. Seed Phytochemistry: resins leaf bark root Phytochemistry: hydrogen cyanide plant Phytochemistry: fish-poisons plant Phytochemistry: insecticides, arachnicides plant Phytochemistry: antibiotic, bacteristatic, fungistatic seed Phytochemistry: fatty acids, etc. leaf root Phytochemistry: alkaloids.

7. *Arctium lappa* (Compositae)

Common names
Eng.: Burdock

Distribution

It grows freely throughout England (through rarely in Scotland) on waste ground and about old buildings, by roadsides and in fairly damp places.

Parts used: Root, herbaloeacemmanana

Pharmacological Activities

Alterative, diuretic and diaphoretic. One of the best blood purifiers. In all skin diseases and potent anticancer.

Chemical Constituents

Inulin, mucilage, sugar, a bitter, crystalline glucoside - Lappin-a little resin, fixed and volatile oils, and some tannic acid.The roots contain starch, and the ashes of the plant, burnt when green, yield carbonate of potash abundantly, and also some nitre.

8. *Astragalus membranaceus* (Papilionaceae)

Common names
Eng.: Green dragon, ogi, radix astragali and yellow leader.
Hindi & Beng.: milk-vetch root

Distribution

This medicinal plant is native to northern and eastern parts of China, as well as Japan,

Mongolia and Korea. It grows along forest margins, on grassy hills and in shrub thickets along hill sides, but it is also found in thin open woods. It thrives in sandy, well-drained soil and full exposure to sunlight.

Parts used: Leaves or herb

Pharmacological Activities

Antiviral, antibacterial, antioxidants, anti-inflammatory, anti-hypertensive, diuretic, nervous conditions, Hodgkin's disease, shortness of breath, persistent infections, fever, some allergies, systemic lupus erythematosus and anemia. Herb's ability to support the body's immune system, particularly in cancer patients.

Chemical Constituents

Swainsonine, saponin, polysaccharides (astragalan I, II and III) and triterpenes.

9. *Asparagus cochinchinensis* (Asparagaceae)

Common names
Eng.: Bamboo Rope
Hindi & Beng.: Tian Men Dong

Distribution

It is found in India E. Asia - China, Japan, Korea.

Parts used: Root

Pharmacological Activities

Antibacterial, Anti-inflammatory, Anti-pyretic, Antiseptic, Anti-tussive, Cancer, Diuretic, Expectorant, Infertility, Nervine, Sialagogue.

Stomachic; Tonic.

Chemical constituents

Spirostanol saponin, asparacoside, spiro-steroids, asparacosins A and B, acetylenic and a new polyphenol and asparenydiol.

10. *Artemisia annua* Linn. (Asteraceae)

Common names
Eng.: Sweet wormwood, sweet annie, sweet fern, sweet sagewort, or annual wormwood.

Distribution

In traditional Chinese medicine, used to treat fever.

Parts used: Leaves

Pharmacological activities

Malaria and cancer and to inhibit a number of viruses, including herpes simplex 1 and hepatitis B and C and cardiovascular.

Chemical constituents

It contains Artemisinin, coumarins, flavones, flavonols, phenolic acids and miscellaneous

11. *Baptisia* (Fabaceae)

Common names
Eng.: Wild indigo

Distribution

It is a perennial indigenous to New England. As it thrives in dry soil, it does not grow around streams or in areas of high precipitation.

Parts used: Herbs

Pharmacological activities

Immunomodulator (it stimulates the immune system). Applied topically, Baptisia soothes inflammation, anticancer and reduces redness. Historically, Baptisia was used as a natural antibiotic in the treatment of both internal and external infections.

Chemical constituents

Baptisia has a high polysaccharide content.

12. *Berberis vulgaris* Linn. (Berberidaceae)

Common names
Oregon grape, Oregon barberry
Eng.: Oregon barberry

Distribution

It is generally found throughout in India. Common barberry and Japanese barberry (*B. thanbergii*), occurs in Europe and North America

Parts used: All Parts

Pharmacological activities

The bark and root bark are antiseptic, astringent, cholagogue, hepatic, purgative, refrigerant, stomachic, tonic, jaundice, general debility and biliousness, antirheumatic, purgative and treatment for diarrhea, diaphoretic. A tincture of the root bark has been used in the treatment of rheumatism, sciatica etc. The root bark is used as anticancer.

Chemical constituents

13. *Betula utilis* (Betulaceae)

Common names
Eng.: Bhojpatra, Birch
Hindi & Beng.: West Himalayan Birch and Bhurja

Distribution

Betul a utilis grows along moraines around Bhojbasa, close to the snout of the Gangotri glacier in India

Parts used: Leaves and bark

Pharmacological activities

Leaves of the plant show efficacy in treatment of urinary tract infections and in kidney and bladder stones. The wood is used for construction, and the foliage for fodder. The most widespread use is for firewood, anti-inflammatory, anti -HIV, antioxidant, anticancer and antimicrobial.

Chemical onstituents

The bark contains betulin, lupeol, oleanolic acid, acetyloleanolic acid, betulic acid, lupenone sitosterol, methylbetulonate, methyl betultriterpenoid, karachic acid.

14. *Blackberry bus* (Rosaceae)

Distribution

Common in Britain is naturalized throughout most of the world, including North America and India.

Parts used: Fruit

Pharmacological activities

Antimicrobial, anticancer, antidysentery, antidiabetic, antidiarrheal, and also good antioxidant.

Chemical constituents

It contains tannins, gallic acid, villosin, and iron; fruit contains vitamin C, niacin (nicotinic acid), pectin, sugars, and anthocyanins and also contains of berries albumin, citric acid, malic acid, pectin and ellagic acid.

15. *Brucea javanica* (L) Merr. (Simaroubaceae)

Common names
Eng.: Fruit makasar, Macassar Kernels

Distribution

Brucea jxavanica grows naturally from Sri Lanka and India to China, Indochina, Malaysia, New Guinea and Australia

Parts used: Fruit

Pharmacological activities

Brucea javanica is mainly used in the treatment of lung and gastrointestinal cancers, antitumoral, antimalarial, and anti-inflammatory properties.

Chemical constituents

Tetracyclic triterpene quassinoids

16. *Coix lacryma-Jobi* Linn. (Gramineae)

Common names
Eng.: Job's Tears
Hindi & Beng.: Gurlu and Gurmur

Distribution

It is found in India and many tropical countries.

Parts used: Root and fruit

Pharmacological activities

It is used as menstrual disorder, blood purifier, antitumorigenic, antioxidant, anti-inflammatory, hepatoprotective, diuretic, anticaner and antipyretic.

Chemical constituents

Leucine, tyrosine, histidine, lysine, arginine and coicin.

17. *Curcuma zedoaria* Roscoe. (Scitaminaceae)

Common names
Eng.: Zedoary, Amomum zerumbet, zerumber, zedoary, white turmeric or kentjur.
Hindi & Beng.: Gandamasti, Kachura and Sutha.

Distribution

Cultivated in gardens in many parts of India, especially in Eastern Bengal and in district of Chittagong and Tipperah.

Parts used: Tubers, leaves and rhizome

Pharmacological activities

Stimulant, Carminative, expectorant, demulcent, diuretic and rubefacient and anticancer. Antiperiodic pills, and antiperiodic tincture. The rhizome is considered to aid digestion, to purify the blood, to provide relief for colic, and for the treatment of colds and infections. The essential oil is an active ingredient in antibacterial preparations. In India the rhizome is chewed to alter a sticky taste in the mouth, and in both Java and India a decoction of the root is used to treat weakness resulting from childbirth.

Chemical constituents

Curcumin, gingerenone A, Gingerols, shogaols and zingerone

18. *Cannabis sativa* Linn. (Cannabaceae)

Common names
Eng.: Marijuana, Indian hemp
Hindi & Beng.: Ganja; Charas and Bhang

Distribution

Plant is native of Persia, Western and Central Asia, now largely cultivated all over India and found wild on Western Himalayas and from Kashmir to east of Assam.

Parts used: All parts

Pharmacological activities

Sedative, anticancer, antibacterial, inflammatory, analgesic and stimulant.

Chemical constituents

Tetrahydrocannabinol (THC), Cannabidiol (CBD), α-Pinene, Myrcene, Linalool, Limonene, Trans-β-ocimene, α-Terpinolene, Trans-caryophyllene, α-Humulene and Caryophyllene-oxide.

19. *Camellia sinensis* (Theaceae)

Common names
Eng.: Tea plant
Hindi & Beng.: Punj, Chha

Distribution

This shrub which is a native of China is grown luxuriantly in the hill districts of India, *viz* Assam, Bengal, (states of Tipperah) Bihar, Orissa, U.P., Punjab, Madras, Coorg, Travancore, Cochin and Mysore.

Parts used: Leaves

Pharmacological activities

Stimulant, diuretic and stringent, headache and anticancer.

Chemical constituents

Epigallocate chin gallate, gallic acid, quercetin, caffeine, xanthine, saponin and theophylline.

20. *Catharanthus roseus* (Apocynaceae)

Common names
Eng.: Vinca
Hindi & Beng.: Sadaphuli and Nayan tara

Distribution

Catharanthus roseus is a fleshy perennial growing to 32 in (80 cm) high. It has glossy, dark green, oval leaves (1-2 inches long) and flowers all summer long. *Catharanthus roseus* is native to the Indian Ocean island of Madagascar. This herb is now common in many tropical and subtropical regions worldwide, including the southern United States.

Parts used: Leaves, whole plant

Pharmacological activities

Asthma and flatulence, entire plant for tuberculosis, dyspepsia, indigestion, diabetes, malaria, high blood pressure. In Africa, leaves are used for menorrhagia and rheumatism and anticancer.

Chemical constituents

Vinblastine, Vincristine, Alstonine, Ajmalicine, Reserpine, Indole alkaloids Vindoline and Catharanthine

21. *Colchicum luteum* Baker (Liliaceae)

Common names
Eng.: Colchicum
Hindi & Beng.: Suranjan, Suranjane-Talkh

Distribution

It is found in India, Pakistan, Middle East and South Africa to Western Europe and Asia

Parts used: Herb

Pharmacological activities

Corms of the *C. luteum* Baker are extensively used for the treatment of gout, rheumatism and diseases of the liver, spleen and blood purifier. It is used as antimicrobial activity, laxative, diuretic and treatment of various cancers.

Chemical constituents

Colchicines, demecolcine, alkaloid, phenol, flavonoids, sterol, tannin and saponins.

22. *Combretum caffrum* Eckl. & Zeyh (Combretaceae)

Common names
Eng.: River bush willow, African bush willow, Cape Bushwillow, Bushveld willow

Distribution

Native to tropical and southern Africa

Parts used: Roots and leaves

Pharmacological activities

It is used for the treatment of a variety of ailments and diseases, ranging from scorpion and snake bites, mental problems, heart and worm remedies to fever and microbial infections and treatment of cancer.

Chemical constituents

Combretastatins (bibenzyl compounds), acidic tetracyclic and pentacyclic triterpenes/ triterpenoids, ellagitannins, phenanthrenes, flavonoids, saponins and cycloartane glycosides.

23. *Curcuma longa* Linn. (Zingiberaceae)

Common names
Eng.: Turmeric Indian saffron, Indian yellow root, Jiang huang.
Hindi & Beng.: Haldi, Haridra,

Distribution

It originates from South or Southeast Asia, from western India.

Parts used: Rhizome and root

Pharmacological activities

Turmeric treats are the skin, heart, liver and lungs. Turmeric is used for epilepsy and bleeding disorders, skin diseases; to purify the body-mind is the most common use of turmeric in ayurveda. Turmeric reduces fevers, diarrhea, urinary disorders, insanity, poisoning, cough, lactation problems and anticancer.

Chemical constituents

Twomerone, curcumine, caryophyllene, chromium, cineole, cinnamic-acid, cobalt, copper, cuminyl-alcohol, curcumene, curcumenol, curcumin, curdione, curlone, curzerenone, curzerenone-c, cyclo-isoprenemyrcene, D-alpha-phellandrene, D-camphene, D-camphor, D-sabinene, Dehydroturmerone, Didesmethoxycurcumin, Di-p-coumaroyl-methane.

24. *Echinacea angustifolia* (Asteraceae)

Common names
Eng.: Black sampson

Hindi & Beng.: Narrow-leaf coneflower, k ansas snakeroot, narrow-leaf purple coneflower

Distribution

Native Americans and by traditional coneflwarin the United States and in Canada.

Parts used: herb

Pharmacological activities

Immune supporting, depurative, vulnerary confeonna, lymphatic, sialagogue and anticancer.

Chemical constituents

Arabinogalactan and fuco galacto-xyloglucans.

25. *Fagopyrum esculentum* Gaertn. (Polygonaceae)

Common names
Eng.: Buck wheat
Hindi & Beng.: Kaspat

Distribution

It is found in central Asia

Parts used: Fresh plant

Pharmacological activities

Itching of the skin, red sore of blotches, pruritus vulvae with yellow leucorrhoea also itching of the knees, elbows and hairy parts and used in the treatment of cancer.

Chemical constituents

Amygdalin, Rutin

26. *Ferula asafoetida* Discuss. (Umbelliferae)

Common names

Eng.: Asafoetida, awei

Hindi & Beng.: Mvuje and devil's dung

Distribution

Ferula is a perennial herb commonly found in Afghanistan mountains but is also cultivated in nearby India.

Parts used: Rhizome and root

Pharmacological activities

Antioxidant, anti-inflammatory and synergistic protective action against oxidative stress in skin and, by extension, photo aging, skin cancer, nervous conditions, bronchitis, asthma and whooping cough, infantile pneumonia and flatulent colic. The gum resin is antispasmodic, carminative, expectorant, laxative, and sedative. The volatile oil in the gum is eliminated through the lungs, making this an excellent treatment for asthma.

Chemical constituents

It contains alpha-pinene and luteolin, two compounds that have anticancer properties also contain ferulic acid.

27. *Ginkgo biloba* (Ginkgoaceae)

Common names
Eng : Kew tree, maidenhair tree
Hindi & Beng.: Bulkawari

Distribution

Ginkgo has long been cultivated in China. It has also been commonly cultivated in North America and in Europe.

Parts used: Leaves

Pharmacological activities

Ginkgo has long and widely been used by the Chinese as a cure for asthma, tuberculosis, chronic coughs and bronchitis. Ginkgo leaves stimulates the heart, improves blood circulation, lung and brain function, treats Alzheimer's disease, varicose veins, Raynaud's disease, hemorrhoids and leg ulcers. The leaves are also helpful in treating cataracts, muscular degeneration, retinopathy, neuropathy, nephropathy, stroke, tinnitus, vertigo, chilblains and impotence. It possesses anti-inflammatory and natural antihistamine and anticancer.

Chemical constituents

Ginkgolide-B, A, C and J. Leaves contain phenolic acids, proanthocyanidins, flavonoid glycosides, such as myricetin, kaempferol, isorhamnetin and quercetin, and the terpenetrilactones, ginkgolides and bilobalides. The leaves also contain unique gingko biflavones, as well as alkylphenols and polyprenols.

28. *Glycine max* Merr. (Leguminosae)

Common names
Eng.: Soyabean

Distribution

It is found in India East Asia, Africa and China.

Parts used: Seed

Pharmacological activities

Breast cancer, cholesterol and heart diseases, antioxidant and phytic acid. The beneficial claims for phytic acid include reducing cancer, minimizing diabetes and reducing inflammation.

Chemical constituents

Zinc, selenium, vitamins (A, B1, B2, B12, C, D, E and K), amino acids, isoflavones, protease inhibitors, saponins, phytosterols, genistein and daidzein

29. *Glycyrrhiza glabra* (Leguminosae)

Common names
Eng.: Liquorice
Hindi & Beng.: Mulethi, Jethimadhu and Yashtimadhu.

Distribution

It is found throughout in India. *Glycyrrhiz aglabra* is native to Eurasia, northern Africa and Western Asia, Russia, Spain and the Middle East. Where it grows up to 1,200 m above sea level.

Parts used: Root, stolon and rhizomes

Pharmacological activities

Anti-allergic, anti-arthritic, anti-inflammatory, demulcent, emollient, estrogenic (mild), expectorant, laxative, pectoral (moderate), soothing, Anti-pyretic anti-infective and anti-cancer therapies. Liquorice is used in the production of cough mixtures and throat lozenges, as well as an ingredient to mask the unpleasant taste of some medicines.

Chemical constituents

Liquorice root contains triterpenoid saponins (4g%, 20%), mostly glycyrrhizin, a mixture of potassium and calcium salts of 18β-glycyrrhizic acid (also known as glycyrrhizic or glycyrrhizinic acid and a glycoside of glycyrrhetinic acid), which is 50 times sweeter than sugar. Other triterpenes present are liquiritic acid, glycyrretol, glabrolide, isoglaborlide and liquorice acid. Various types, including flavanones or flavanonols, chalcones, isoflavans, isoflavenes, flavones or flavonols, isoflavones and isoflavanones. Amongst them, flavanones and chalcones are the main types.

30. *Gossypium barbadense* Linn. (Malvaceae)

Common names
Eng.: Raw cotton
Hindi & Beng.: Basri

Distribution

Perennial under shrub 1–3 m high, found in India and growing in Sind, Assam and U.P. Native of NW S America, West Indies, Mexico, North-East Africa, Arabia and Australia.

Parts used: Leaves, seed

Pharmacological Activities

Traditional medicine *Gossypiumbarbadense* L. is used against hypertension, antimicrobial, insecticidal and cytotoxic and anti-neoplastic agent.

Chemical constituents

Gossypol, triterpenoid and sesquiterpenoid aldehydes compounds, flavonols, tannins, cytotoxic sulfated compounds and sesquiterpene glycosides.

31. *Gyrophora esculenta* Miyoshi. (Umbilicariaceae)

Common names
Eng.: Mushroom

Distribution

It is grows on rocks. It can be found in East Asia including in China, Japan, and Korea. It is edible when properly prepared and has been used as a food source and medicine.

Parts used: Whole part

Pharmacological activities

Inhibits growth of cancer by enhancing activity of the natural killer cells. A study revealed that it inhibits carcinogenesis and metastases.

Chemical VBonstituents

Polysaccharides β-glucans, α-glucans, galactomannans, lecanoric and gyrophoric acids.

32. *Helianthus annuus* Linn. (Asteraceae or Compositae)

Common names
Eng.: Sunflower
Hindi & Beng.: Hurduja; Suraj-mukhi and Surajmukhi

Distribution

This plant is common in lIndian gardens and

also North America, Southern Canada, and Mexico at elevations below 1900 m. *Helianthus annuus* is highly variable as a species, and hybridizes with several other species. The heads and plants are very large in cultivated forms.

Parts used: Seed oil, flower

Pharmacological activities

Herbicides, insecticides, fungicides, anticancer and malaria.

Chemical constituents

Unsaturated fatty acid, alkaloids, amino acis.

33. *Lentinus edodes* (Agaricaceae)

Common names
Eng.: Mushrooms
Hindi & Beng.: Mushrooms

Distribution

It is found China and Japan

Parts used: Whole part

Pharmacological activities

Nutritional and medicinal properties, antibiotic, anti-carcinogenic and antiviral compounds.

Chemical Constituents

Lentinan, lentins, lectins and eritadenins.

34. *Linumusitatissimum* Linn. (Linaceae)

Common names
Eng.: Flaxseed, Linseed
Hindi & Beng.: Tisi; Alsi and Masina

Distribution

Flax plant is a native of Egyptian, extensively cultivated in India, chiefly in Bengal, Bihar and United Provinces. It has been cultivated in all temperate and tropical regions for so many centuries, escaping from cultivation and naturalizing, that it is found semi-wild in almost all countries in which it is cultivated.

Parts used: Seed, oil and flower

Pharmacological activities

It is commonly used as a stimulant, diaphoretic, expectorant, diuretic, antioxidant properties cancers.

Chemical constituents

Cynogenetic glycosides, Lignans, linoleic acid, amygdalin, resin, Linoxyn, stearic acid palmitic acid, myristic acid and linamarin.

35. *Mentha species* (Labiatae)

Common names
Eng.: Pudina

Distribution

It is found in India, Australia, South Africa and North America.

Parts used: Leaves

Pharmacological Activities

It is commonly used as a stimulant, aromatic, carminative, expectorant and cancers.

Chemical Constituents

Monoterpene ketones, alkaloids, flavonoids, phenols, gummy polysaccharides. Terpens and quinons are used in food and pharmaceutical, cosmetics and pesticide industries.

36. *Momordica charantia* Linn. (Cucurbitaceae)

Common names
Bitter melon, bitter gourd, bitter squash, patrick or balsam-pear
Eng.: Bitter Melon
Hindi & Beng.: Karela

Distribution

It is a tropical and subtropical vine of the family Cucurbitaceae, widely grown in Asia, Africa, and the Caribbean for its edible fruit,which is extremely bitter.

Parts Used: Fruit

Pharmacological Activities

Diabetes, as a stomachic, laxative, antibilious, emetic, anthelmintic agent, for the treatment of cough, respiratory diseases, skin diseases, wounds, ulcer, gout, rheumatism and anticancer properties.

Chemical Constituents

These includes proteids, triterpenes, lipids, inorganic compounds, phenylpropanoids, carotenoids, steroids, alkaloids, monoterpenes, alkene to C3, carbohydrates, benzanoids, alkanol C5 or more, other unknown structure (e.g., kakara I-B, II-A and III-B) sterol and sesquiterpene. Of the 228 different compounds, most of these fall under the groups of proteids and triterpenes.

37. *Nothapodytes foetida* (Wight) Baehni. (Simaroubaceae)

Common names
Eng.: Nothapodytes Tree

Distribution

It is found in India-cultivated in Karnataka: Chikmagalur, Coorg, Hassan, N. Kanara, Shimoga.

Parts used: Wood of tree

Pharmacological Activities

Anticancers

Chemical Constituents

Acetylcamptothecin, Camptothecin, Scopolectin

38. *Oroxylum indicum* Linn. (Bignoniaceae)

Common names
Eng.: Midnight horror, oroxylum, kampong, Indian trumpet flower
Hindi & Beng.: Kutannat, soran and Sona

Distribution

Oroxylumindicum is native to the Indian subcontinent, in the Himalayan to Bhutan and southern China, in Indo-China and the Maleysia ecozone. It is visible in the forest biome of Manas National Park in Assam, India. It is found, in the forest areas of the Banswara district in the state of Rajasthan in India. It is reported in the list of rare, endangered and threatened plants of Kerala (South India).

Parts used: Seed and root bark

Pharmacological Activities

Anti-inflammatory, anti-allergy effects, rheumatism and anticancer.

Chemical Constituents

Chrysin, baicalein, Tetuin, 6-glucoside of baicalein, flavonoids.

39. ***Ochrosia elliptica*** **Labill. (Apocynaceae)**

Common names
Eng.: elliptic yellow wood or kopsia, Bloodhorn, Mangrove ochrosia, Wedge apple.
Hindi & Beng.: Berrywood tree

Distribution

North eastern Australia and western Pacific. Introduced in many subtropical and topical regions of the world.

Parts Used: Leaves, bark and flower

Pharmacological Activities

Potent anticancer agent and its derivatives are used to treat cancers of the breast and the kidney. Ornamental (although fruits highly poisonous). Leaf and stem material of this species used against some tumors. This plant was used in colonial medicine; bark used to treat malaria, but contains no quinine.

Chemical Constituents

Ellipticine and 9-methoxy ellipticine are pyridocarbazole (monomeric indole) alkaloids.

40. ***Panax ginseng*** **(Aralaceae)**

Common names
Eng.: Ginseng
Hindi & Beng.: Plant root

Distribution

Ginseng is found in North America and in eastern Asia (mostly northeast China, Korea, Bhutan, eastern Siberia), typically in cooler climates.

Parts Used: Root

Pharmacological Activities

Ginseng seemed to be most protective against cancer of the ovaries, larynx, pancreas, esophagus, and stomach and less effective against breast, cervical, bladder, and thyroid cancers. It is also used as mild stomachic tonic and stimulant, useful in loss of appetite and in digestive affections that arise from mental and nervous exhaustion.

Chemical Constituents

Ginsenosides, Panaxosides, 6 triterpene saponins called ginsenosides. Other active constituents include flavonoids, polysaccharides and polyacetylenes, essential oils, phytosterols, amino acids, peptides, vitamins and minerals.

41. ***Picrorhiza kurroa*** **Benth. (Scrophulariaceae)**

Common names
Eng.: Picrorhiza (kutki), Katuka, Kutka
Hindi & Beng.: Kutki and Kuru

Distribution

It is found in India - Cultivated mainly in North Western Himalayas from Kashmir to Sikkim.

Parts Used: Rhizomes, root

Pharmacological activities

It helps in curing loss of appetite, liver diseases, jaundice, bronchial asthma, bile disorders and weakness. In larger doses, it is useful as an antioxidant, antibacterial, anticholestatic, antiallergic, antiperiodic, cathartic, laxative, and to treat abdominal diseases, adiposity and related disorders and anticancer.

Chemical constituents

Picrosides I, II, III and kutkoside. The root of this plant is rich in glucosides including kitkin and picrorhizin, cucurbitacin, D-mannitol, benetic acid, kutkisterol, vanillic acid and some steroids. P. kurroa also contains apocynin.

42. *Podophyllum hexandrum* (Berberidaceae)

Common names
Eng.: Podophyllum, Himalayan Mayapple, Indian mayapple.
Hindi & Beng.: Ban kakari and papra.

Distribution

Himalayan Mayapple is a perennial herb, 15-40 cm tall, native to the Himalayas. It is low to the ground with glossy green, drooping, lobed leaves on its few stiff branches.

Parts used: Root, stem

Pharmacological activities

Constipation, cold, bacterial infections, biliary fever, sceptic wounds and insect bites, allergic and inflammatory conditions. The root and underground stem (rhizome) are used to make medicine. Himalayan mayapple is used orally for jaundice, liver ailments, fever, syphilis, hearing loss and cancer.

Chemical constituents

Podophyllin, astragalin

43. *Phaleria macrocarpa* Buah. (Thymelaeaceae)

Common names
Eng.: God's Crown, Mahkota Dewa

Distribution

It is a dense evergreen tree, indigenous to Indonesia and Malaysia. It is found in tropical areas of New Guinea up to 1,200 metres above sea level.

Parts used: All part

Pharmacological Activities

It is used as anti-hyperglycemia, anti-tumor, anti-inflammation, anti-diarrhoeal, anti-oxidant, anti-viral, anti-bacterial and anti-fungal, vasodilator. The stem and shells of seeds have been used to treat various cancers, lung, live and heart diseases. The leaves also contain compounds that can treat impotence, blood diseases, allergies, diabetes mellitus and tumors, antihyper cholesterolemia activity and antiatherosclerosis.

Chemical constituents

Phalerin, gallic acid, Icaricide C, mangiferin, mahkoside A, dodecanoic acid, palmitic acid, des-acetylflavicordin-A, flavicordin-A, flavicordin-D, flavicordin-A glucoside, ethyl stearate, lignans, alkaloids and saponins.

44. *Smilax china* (Liliaceae)

Common names
Eng.: Madhunuhi
Hindi & Beng.: Topchini, kumarika
Sans.: Madhusnuhi

Distribution

It is native to China, Korea, Taiwan, Japan (including Ryukyu and Bonin Islands), Philippines, Vietnam, Thailand, Myanmar, and Assam.

Parts used: Root, rhizomes

Pharmacological activities

The rhizomes are bitter, acrid, thermogenic, anodyne, anti-inflammatory, digestive, laxative, depurative, diuretic, febrifuge and tonic. It is used in dyspepsia, flatulence, colic, constipation and helminthiasis. It is useful in skin diseases, leprosy and psoriasis. It is used in fever, epilepsy, insanity, neuralgia and anticancer. It is used syphilis, strangury, seminal weakness and general debility. Detoxifies organs, cleanses blood, aids absorption and kills bacteria. It also stimulates digestion, increases urination, protects liver and promotes perspiration.

Chemical constituents

Kaempferol 7-O-glucoside, a flavonol glycoside

45. *Solanum nigrum* Linn. (Solanaceae)

Common names
Eng.: Leunca
Hindi & Beng.: Makoi; Gurkamai and Gurkamai; Tulidun

Distribution

It is common herb throughout India. These species are only semi-cultivated in a few countries in Africa and Indonesia, cultivation of the garden huckleberry for its fruits in North America.

Parts used: Leaves

Pharmacological activities

Exhibiting anti-tumor activity, diuretic, expectorant, sedative, anti-neoplastic activity, hepatoprotective and antioxidant.

Chemical constituents

Alkaloids, flavonoids, tannins, saponins, glycosides, proteins, carbohydrates, coumarins and phytosterols. It has been found that *Solanum nigrum* contains the substances, such as total alkaloid, steroid alkaloid, steroidal saponins glycoprotein, gallic acid, PCA, catechin, caffeic acid, epicatechin, rutin and narigenin.

46. *Typhonium flagelliforme* (Lodd) Blume. (Araceae)

Common names
Eng.: Rodent Tuber
Hindi & Beng.: Keladi, Tikus

Distribution

It is found in India - Cultivated mainly in Karnataka, Tamil Nadu dist, Kottayam, Alappuzha, Kasaragod, Kollam, Pathanamthitta, Kannur, Thiruvanantha-puram, Kozhikode

Parts used: Leaves, stem and rhizome

Pharmacological activities

Malaysia, is often used as an essential ingredient of herbal remedies for alternative cancer therapies. It can detoxicate, reduce swelling, get rid of boils and pus, stop bleeding, ease pain and is effective in the treatment of lymphatic problems. It is also use to increase one's appetite, improve health, revitalize the person who would tire easily.

Chemical constituents

Methyl esters of hexadecanoic acid, octadecanoic acid, 9-octadecenoic acid and 9, 12-octadecadienoic acid. In addition, several common aliphatics were identified as dodecane, tridecane, tetradecane, pentadecane, hexadecane, heptadecane, octadecane, nonadecane and eicosane. The unique methyl ester of 13-phenyltridecanoic acid.

47. *Taraxacum mongolicum* (Asteraceae)

Common names

Eng.: Handsome

Distribution

Asia, Europe, and North America

Parts used: Leaves and root

Pharmacological activities

Root is used as gastrointestinal remedy supporting digestion and liver function, while the leaf is used as a diuretic and bitter digestive stimulant, inflammation modulator, diuretic, digestive stimulant, insulin stimulant, demulcent, prebiotic, immunomodulator, antiangiogenic, and antineoplastic.

Chemical constituents

Quercetin, luteolin, luteolin-7-O-beta-D-glucopyranoside, caffeic acid, esculetin, stigmasterol and taraxasterol acetate.

48. *Taxus brevifolia* (Taxaceae)

Common names

Eng.: Pacific yew

Distribution

Native tribes in western North America long used Pacific yew for its healing powers. Pacific yew is a very common tree or shrub in the coniferous forest understory of western North America, along the coastal ranges from southeastern Alaska to northern California and in the southern Canadian Rockies to central Idaho and neighboring Montana.

Parts used: All parts of the plant

Pharmacological activities

Pharmacologists have conducted conclusive studies to show that taxol, a complex molecule in Pacific yew, found in all parts of the plant, is effective for chemotherapy of ovarian and possibly other cancers.

Chemical constituents

Taxanes, taxol cepholomannine

49. *Tabebuia impetiginosa, T. avellanedae* (Bignoniaceae)

Common names

Eng.: Lapacho, Pau D'Arco, Taheebo, and IpeRoxo

Distribution

It is found in India.

Parts used: Bark, wood

Pharmacological activities

Anti-cancer properties that are being researched with focus on pancreatic cancer.

Antibacterial, inhances immunity, mildly laxative, relieves rheumatism, relieves pain, reduces inflammation and reduces tumors.

Chemical constituents

Beta-Lapachone, Lapachol Beta-lapachone

50. *Withania somnifera* Dunal. (Solanaceae)

Common names
Eng.: Ashwagandha, Indian ginseng, poison gooseberry or winter cherry.

Hindi & Beng.: Ashwagandh and Ashwagandha

Distribution

It is shrub is common in Bombay and Western India, occasionally met with in Bengal. *Withania somnifera* is cultivated in many of the drier regions of India, such as Mandsaur District of Madhya Pradesh, Punjab, Sindh, Gujarat, and Rajasthan. It is also found in Nepal, China and Yemen.

Parts used: Leaves and root

Pharmacological activities

In Ayurveda, the berries and leaves are applied externally to tumors, tubercular glands, carbuncles, and ulcers, nervine sedative, astringent, burns and wounds, as also for a sunscreen upon women's faces.

Chemical constituents

Withanolides, Withaferin

51. *Zingiber officinale* Roscoe. (Zingiberaceae)

Common names
Eng.: Ginger
Hindi & Beng.: Adrak; Duk and Sonth

Distribution

Ginger is said to be a native of China and India. It is cultivated in West Indies, Jamaica, Africa.

Parts used: Rhizome

Pharmacological Activities

Ginger root helps relieve the dizziness, sweating, nausea, and vomiting that comes from motion sickness or seasickness. It can also ease sore throats, headaches, ulcerative colitis, anticancer some types of menstrual and arthritis pain, and fevers and aches caused by colds and flu. It may also help to relieve symptoms of depression.

Chemical Constituents

The active chemicals in the plant, such as zingiberene, bisabolene, gingerol, Shogaol, Curcumin, gingerenone A and Gingerol.

9

MEDICINAL PLANTS FOR ANTIMICROBIAL

INTRODUCTION

The discovery and development of antibiotics are among the most powerful and successful achievements of modern science and technology for the control of infectious diseases. However, the rate of resistance of pathogenic microorganisms to conventionally used antimicrobial agents is increasing with an alarming frequency. Isolation of microbial agents less susceptible to regular antibiotics and recovery of resistant isolates during anti bacterial therapy is increasing throughout the world. In addition to this problem antibiotics are some times associated with adverse side effects on the host, which include hypersensitivity, depletion of beneficial gut and mucosal microorganisms, immune suppression and allergic reactions. The number of multi-drug resistant microbial strains and the appearance of strains with reduced susceptibility to antibiotics are continuously increasing. This increase has been attributed to indiscriminate use of broad-spectrum antibiotics, immunosuppressive agent, intravenous catheters, organ transplantation and ongoing epidemics of HIV infection. Examples include methicillin-resistant staphylococci, pneumococci resistant to penicillin and macrolides, vancomycin-resistant Enterococci as well as multi-drug resistant gram-negative organisms. There is an urgent need to control antimicrobial resistance by improved antibiotic usage and reduction of hospital cross-infection, however, the development of new antibiotics should be continued as they are of primary importance to maintain the effectiveness of antimicrobial treatment. The potential for developing antimicrobials from higher plants appears rewarding as it will lead to the development of a phytomedicine to act against microbes; as a result, plants are one of the bed rocks for modern medicine to attain new principles. Plant based antimicrobials represent a vast untapped source of medicine. Plant based antimicrobials have enormous therapeutic potential as they can serve the purpose without any side effects that are often associated with synthetic antimicrobials. Further continued exploration of plant derived antimicrobials is needed today. Historically, plants have provided a source of inspiration for novel drug compounds, as plant derived medicines have made large contributions to human health and well-being. Medicinal plants constitute an effective source of both traditional and modern medicines. Herbal medicine has been shown to have genuine utility and about 80% of rural population depends on it as primary health care. Over the years, the World Health Organization advocated that countries should encourage traditional medicine with a view to identifying and exploiting aspects that provide safe and effective remedies for ailments of both microbial and non-microbial origins. In recent years, pharmaceutical companies have spent a lot of time and money in developing natural products extracted from plants, to produce more cost effective remedies that are affordable to the population. Scientific experiments on the antimicrobial properties of plant components were first documented in the late 19th century. It is estimated that today, plant materials are present in, or have provided the models for 50% Western drugs. Many commercially proven drugs used in modern medicine were initially used in crude form in traditional or folk healing practices, or for other purposes that suggested potentially useful biological activity. The primary benefits of using plant derived medicines are that they are relatively safer than synthetic alternatives, offering profound therapeutic benefits and more affordable treatment.

The medicinal plants around the world contain many compounds with antibacterial activity. Many efforts have been made to discover new antimicrobial compounds from various sources such as micro-organisms, animals, and plants. Systematic screening of them may result in the discovery of novel effective antimicrobial compounds. The use of botanical medicines is generally on the rise in many parts of the world. The screening of plant extracts and plant products for antimicrobial activity has shown that plants represent a potential source of new anti-infective agents. Numerous experiments have been carried out to screen natural products for antimicrobial property. Considering the above, it can be stated that plants are valuable sources for new compounds and should receive special attention in research strategies to develop new antimicrobials urgently required in the near future.

The science dealing with the study of the prevention and treatment of diseases caused by micro-organisms is known as medical microbiology. Its subdisciplines are virology (study of viruses), bacteriology (study of bacteria), mycology (study of fungi), phycology (study of algae) and protozoology (study of protozoa). For the treatment of diseases inhibitory chemicals employed to kill micro-organisms or prevent their growth, are called antimicrobial agents. These are classified according to their application and spectrum of activity, as germicides that kill micro-organisms, whereas micro-biostatic agents inhibit the growth of pathogens and enable the leucocytes and other defense mechanism of the host tocope up with static invaders. The germicides may exhibit selective toxicity depending on their spectrum of activity. They may act as viricides (killing viruses), bacteriocides (killing bacteria), algicides (killing algae) or fungicides (killing fungi).

The beginning of modern chemotherapy has largely been due to the efforts of Dr. Paul Ehrlich (1910), who used salvarsan, as arsenic derivative effective against syphilis. Paul Ehrlich used the term chemotherapy for curing the infectious disease without injury to the host's tissue, known as chemo-therapeutic agents such as antibacterial, antiprotozoal, antiviral, antineoplastic, antitubercular and antifungal agents. Later on, Domagk (1953) prepared an important chemotherapeutic agent sulfanilamide.

Classification of Antibacterial Agents

The antibacterial agents are classified in three categories:

(I) Antibiotics and chemically synthesized chemotherapeutic agents.

(II) Non-antibiotic chemotherapeutic agents (Disinfectants, antiseptics and preservatives)
(III) Immunological products.

(I) *Antibiotics*

They are produced by micro-organisms or they might be fully or partly prepared by chemical synthesis. They inhibit the growth of micro-organisms in minimal concentrations. Antibiotics may be of microbial origin or purely synthetic or semi synthetic. They can be classified by manner of biosynthesis or chemical structure. Structurally, they are classified into different classes as shown in the following table.

Classification of Antibiotics According to Their Chemical Structure

1. Carbohydrate-containing antibiotics
2. Macrocyclic lactones
3. Quinones and related antibiotics
4. Amino acid and peptide antibiotics
5. Heterocyclic antibiotics containing Oxygen
6. Heterocyclic antibiotics containing nitrogen
7. Aromatic antibiotics
8. Aliphatic antibiotics

Synthetic antimicrobial agents include sulfonamides, diamino pyrimidine derivatives, antitubercular compounds, nitrofuran compounds, 4-quinoline antibacterials, imidazole derivatives, flucytosine etc.

(II) *Non-antibiotics*

The second category of antibacterial agents includes non-antibiotic chemotherapeutic agents which are as follows:

1. *Acids and their derivatives*

Some organic acids such as sorbic, benzoic, lactic and propionic acids are used for preserving food and pharmaceuticals. Salicyclic acid has strong antiseptic and germicidal properties as it is a carboxylated phenol. The presence of –COOH group appears to enhance the antiseptic property and to decrease destructive effect. Benzoic acid is used externally as an antiseptic and is employed in lotion and ointment. Benzoic acid and salicylic acid are used to control fungi that cause disease such as athlete's foot. Benzoic acid and sodium benzoate are used as antifungal preservatives. Mandolic acid possesses good bacteriostatic and bactericidal properties.

2. *Alcohols and related compounds*

They are bactericidal and fungicidal, but are not effective against endospores and some viruses. Various alcohols and their derivatives have been used as antiseptics *e.g.*, ethanol and propanol. The antibacterial value of straight chain alcohols increases with an increase in the molecular weight and beyond C8 the activity begins to fall off. The isomeric alcohol shows a drop in activity from primary, secondary to tertiary. Ethanol has extremely numerous uses in pharmacy.

3. *Chlorination and compound containing chlorine*

Chlorination is extensively used to disinfect drinking water, swimming pools and for the treatment of effluent from industries. Robert Koch in 1981 first referred to the bactericidal properties of hypo-chlorites. N-chloro compounds are represented by amides, imides and amidines where in one or more hydrogen atoms are replaced by chlorine.

4. *Iodine containing compounds*

Iodine containing compounds are widely used as antiseptic, fungicide and amoebicide. Iodophores are used as disinfectants and antiseptics. The soaps used for surgical scrubs often contain iodophors.

5. *Heavy metals*

Heavy metals such as silver, copper, mercury and zinc have antimicrobial properties and are used in disinfectant and antiseptic formulations. Mercurochrome and merthiolate are applied to skin after minor wounds. Zinc issused in antifungal antiseptics. Copper sulfate is used as algicides.

6. *Oxidizing agents*

Their value as antiseptics depends on the liberation of oxygen and all are organic compounds.

7. *Dyes*

Organic dyes have been extensively used as antibacterial agents. Their medical significance was first recognized by Churchman in 1912. He reported inhibitory effect of Crystal violet on Gram-positive organism. The acridines exert bactericidal and bacteriostatic action against both Gram-positive and Gram- negative organisms.

8. *8-Hydroxyquinolines*

8-Hydroxyquinoline or oxine is unique among the isomeric hydroxyquinolines, for it alone exhibits antimicrobial activity. This attributes to its abilityto chelate metals, which the other isomers do not exhibit.

9. *Surface active agents*

Soaps and detergents are used to remove microbes mechanically from the skin surface. Anionic detergents remove microbes mechanically; cationic detergents have antimicrobial activities and can be used as disinfectants and antiseptics.

(III) *Immunological products*

Certain immunological products such as vaccines and monoclonal antibodies are used to control the diseases as a prophylactic measure.

Mode of Action

Antimicrobial drugs interfere chemically with the synthesis of function of vital components of microorganisms. The cellular structure and functions of eukaryotic cells of the human body. These differences provide us with selective toxicity of chemotherapeutic agents against bacteria. Antimicrobial drugs may either kill microorganisms outright or simply prevent their growth. There are various ways in which these agents exhibit their antimicrobial activity. They may inhibit

1. Cell-wall synthesis
2. Protein synthesis
3. Nucleic acid synthesis
4. Enzymatic activity
5. Folate metabolism or
6. Damage cytoplasmic membrane

Bacteriostatic dyes

Stearn and Stearn attributed the bacteriostatic activity to triphenylmethane dyes. Fischer and Munzo have found the relationship between their structure and effectiveness of such dyes. A number of drugs are metal-binding agents. The chelates are the active form of drugs. The site of action within the cell or on the cell surface has not been established. The site of action of oxine and its analogs has been suggested inside the bacterial cell or on cell surface.

Detoxification of antibacterial

P-Aminobenzoic acid is a growth factor for certain micro-organisms and competitively inhibits the bacteriostatic action of sulfonamides. The metabolites identified in man are p-amino-benzoylglucoronide; p-aminohippuric acid, p-acetylaminobenzoic acid. 8-Hydroxyquinoline (oxine) and 4-hydroxyquinoline are excreted as sulfate esters or glucuronide.

Bacteria

The bacteria are microscopic organisms with relatively simple and primitive forms of prokaryotic type. Danish Physician Christian Grams, discovered the differential staining technique known as Gram staining, which differentiates the bacteria into two groups "Gram positive" and "Gram negative", Gram positive bacteria retain the crystal violet and resist decolonization with acetone or alcohol and hence appear deep violet in color; while Gram negative bacteria, which loose the crystal violet, are counter-stained by safranin and hence appear red in colour.

These two groups of bacteria are recently classified into four different categories as follows:

1. The world of bacteria I: "Ordinary" Gram negative bacteria.
2. The world of bacteria II: "Ordinary" Gram positive bacteria.
3. The world of bacteria III: "Bacteria" with unusual properties.
4. The world of bacteria IV: Gram positive filamentous bacteria of complex morphology.

PLANTS FOR ANTIMICROBIAL

1. *Acacia auriculiformis* Benth. (Fabaceae)

Common names

Eng.: Auri, earleaf acacia, earpod wattle, northern black wattle, papuan wattle.
Hindi & Beng.: Tan wattle, akashmoni

Distribution

It is native to Australia, Indonesia, and Papua New Guinea. It grows up to 30 m tall.

Parts used

Pharmacological Activities

Antihelmintic, anti-filarial and micro-bicidal activity. Root is known to be useful in the treatment of various aches, pains and sore eyes.

Chemical constituents

Flavonoids and tannin.

2. *Achillemille folium* Linn. (Asteraceae)

Common names

Eng.: Yarrow
Hindi & Beng.: Rojmari

Distribution

It is found in Himalaya from Kashmir to Kumaon.

Parts used: Flower-heads, leaves.

Pharmacological activities

Yarrow flowers have a number of different uses. As a powder, they stop bleeding quickly. Infused in water, it speeds the healing of canker sores. As a tea, yarrow is used to fight urinary tract infections. Because it can cause uterine contractions, avoid during pregnancy.

Chemical constituents

Flavonoids, tannins, coumarins and proazulene.

3. ***Aegle marmelos*** **Corr. (Rutaceae)**

Common names
Eng.: Bael fruit, Bengal quince
Hindi & Beng.: Bel, Baelsripal and Bela, Bael

Distribution

It is found all over India, from sub Himalaya forests, Bengal, Central and South India, and in Burma.

Parts used: Fruits (both ripe and unripe), root- bark, leaves, rind of the ripe fruit and flower.

Pharmacological activities

Antimicrobial, fungicidal, antibacterial, antiviral, analgesic, anti-inflammatory, antitumor, cytotoxic, immunostimulant, antihelminthic, expectorant and antitussive activities.

Chemical Constituents

Triterpene, pectin, sugar, mucilage, tannin, volatile oil, bitter principle and saponin.

4. ***Allium cepa*** **Linn. (Liliaceae)**

Common names
Eng.: Onion.
Hindi & Beng.: Piyaz and Piyaj, Piyang, Pyaj, Pulantic

Distribution

The onion is believed to have been domesticated in central Asia or cultivated all over India.

Parts used: Bulb and seed.

Pharmacological activities

Onion is used as an antimicrobial, cardiovascular-supportive, hypoglycemic, antioxidant/anticancer, and asthma-protective agent. However, few clinical trials are available to support the use of onion for any indication. In folk medicine, onion has been used for asthma, bronchitis, whooping cough, and similar ailments. Other uses include the treatment of stingray wounds, warts, acne, appetite loss, urinary tract disorders, and indigestion. Onion skin dye has been used as an egg and cloth coloring.

Chemical constituents

Onions contain 89% water, 1.5% protein, and vitamins, including B_1, B_2, and C, along with potassium. Polysaccharides such as fructos, saccharose, and others are also present, as are peptides, flavonoids, and essential oil. Onion contains alliin and similar sulfur compounds, including allylalliin and methyl and propyl compounds of cysteine sulfoxide. Sulfur and other compounds of *A. cepa* have been analyzed. Prostaglandins also have been identified in onion.

5. *Allium sativum* Linn. (Liliaceae)

Common names

Eng.: Garlic, allium, stinking rose, rustic treacle, nectar of the gods, camphor of the poor, poor man's treacle
Hindi & Beng.: Lasan and rasun.

Distribution

It is cultivated all over India.

Parts used: Bulb.

Pharmacological activities

Evidence suggests that garlic may beneficially affect cholesterol and lipids. Among its traditional uses, it has been employed for its antiseptic and antibacterial properties. Other potential areas of use include GI disorders and oncology.

Chemical constituents

Fresh garlic is a source of numerous vitamins, minerals, and trace elements.It contains the highest sulfur content of any member of the genus *Allium*. Two trace elements, germanium and selenium, are found in detectable quantities and have been postulated to play a role in the herb's antitumor effect. Garlic contains about 0.5% of a volatile oil composed of sulfur-containing compounds (diallylisulfide, diallyl trisulfide, methyl allyl trisulfide). The bulbs contain an odorless, colorless, sulfur-containing amino acid called alliin (S-allyl-L-cysteine sulfoxide), which has no known pharmacologic activity. When the bulb is ground, the enzyme allinase is released, resulting in the conversion of alliin to 2-propenesulfenic acid, which dimerizes to form allicin. Allicin gives the pungent characteristic odor to crushed garlic and is believed to be responsible for some of the pharmacologic activity of the plant.

6. *Arctostaphylos uva-ursi* Spreng. (Ericaceae)

Common names

Eng.: *Uva ursi* and pinemat Manzanita.
Hindi & Beng.: Kinnikinnick.

Distribution

Distribution in northern North America, Asia and Europe.

Parts used:

Pharmacological activities

Uva ursi is often used as a tincture or capsule for treating urinary tract infections as it contains compounds effective at killing pathogens typically associated with UTIs.

Chemical constituents

Arctostaphylos uva-ursi contains the glycoside arbutin. Some constituents such as the hydroquinones are hepatotoxic and in cases of urinary tract infections.

7. *Cassia auriculata* Linn. (Caespiniaceae)

Common names

Eng.: Mature tea tree, Tanner's cassia.
Hindi & Beng.: Tarwar.

Distribution

It grows wild in the central Provinces, Western Coast, South India and Ceylon.

Parts used: Root, leaves, flower, bark and seeds.

Pharmacological Activities

Antimicrobial, fungicidal, antibacterial, antiviral, analgesic, anti-inflammatory, antitumor, cytotoxic, immunostimulant, anhelmintic, expectorant, astringent and antitussive activities.

Chemical Constituents

Triterpene, tannin and saponin.

8. *Cinnamomum zeylanicum* Breyn. (Lauraceae)

Common names

Eng.: Cinnamon, cinnamomum, ceylon cinnamon, Chinese cinnamon.

Hindi & Beng.: Punj and Dalchini, Daruchini.

Distribution

The plant is native to Sri Lanka, Southeastern India, Indonesia, South America and the West Indies.

Parts used: Bark

Pharmacological activities

Cinnamon used as a spice and an aromatic. The bark or oil has been used to combat microorganisms, diarrhea and other GI disorders, and dysmenorrhea. Research interest has focused on cinnamon's potential as an insulin-like analog, an anti-inflammatory agent, an antioxidant, and an antimicrobial substance.

Chemical Constituents

The essential oil is cinnamaldehyde and lesser amounts of other phenols and terpenes, including eugenol, trans-cinnamic acid, hydroxycinnamaldehyde, o-methoxycinnamaldehyde, cinnamylalcohol and its acetate, limonene, α-terpineol, tannins, mucilage, oligomericprocyanidins, and trace amounts of coumarin.

9. *Gaultheria procumbens* Linn. (Ericaceae)

Common names

Eng.: Wintergreen, teaberry, checkerberry, gaultheria oil, boxberry, deerberry, Canada tea, partridgeberry.
Hindi & Beng.: Mountain tea.

Distribution

Wintergreen is a perennial evergreen shrub with thin, creeping stems from which leathery leaves with toothed, bristly margins arise. It is a low-growing plant native to eastern North America and usually is found in woodland and exposed mountainous areas.

Parts used: Leaves and fruits.

Pharmacological activities

American Indians reportedly used wintergreen for treating back pain, rheumatism, fever, headaches, antimicrobial and sore throats. The plant and its oil have been used in traditional medicine as an anodyne, analgesic, carminative, astringent, and topical rubefacient.

Chemical Constituents

Methyl ester, methyl salicylate and gaultherin.

10. *Melissa officinalis* Linn. (Lamiaceae)

Common names
Eng.: Lemon balm, balm, common balm or balm mint

Distribution

Native to south-central Europe, North Africa, the Mediterranean region and Central Asia.

Parts used: Leaves.

Pharmacological Activities

Anti-bacterial, anti-oxidant, anti-spasmodic, anti-viral, aromatic, carminative, cerebral stimulant, diaphoretic, digestive, emmenagogue, febrifuge, nervous restorative, spasmolytic, sedative (mild) and tonic.

Chemical Constituents

Essential oils (containing citral and citronellal monoterpenes), flavonoids and rosmarinic, caffeic and chlorogenic acids.

11. *Morinda citrifolia* Linn. (Rubiaceae)

Common names
Eng.: Morinda, noni, hog apple, Indian mulberry, mengkudu, morade la India, pain killer, ruibarbocaribe, wild pine.
Hindi & Beng.: Al, Achi and Ach.

Distribution

The morinda plant, native to Asia, Australia and Polynesia, is a 3 to 8 m high tree or shrub.

Parts used: Fruits.

Pharmacological activities

Morinda has been used for heart remedies, arthritis, headache, digestive and liver ailments, diabetes, high blood pressure, arthritis and aging.

Chemical constituents

Morinda citrifolia fruits contain essential oils with hexcanoic and octonic acids, paraffin and esters of ethyl and methyl alcohols. Fresh plants contain anthraquinones, morindone and alizarin. A new anthraquinone glycoside from morinda heart wood has recently been described.

12. *Myristica fragrans* Houtt. (Labiatae)

Common names
Eng.: Nutmeg, mace, magic, muscdier, nux moschata, myristica oil, muskatbaum.
Hindi & Beng.: Jayphal, Jaiphal, Jaepatri.

Distribution

Cultivated in India, Ceylon, Malaysia, and Granada. The fruit, which is called a drupe or a nutmeg apple, is similar in appearance to a peach or an apricot.

Parts used: Fruits, seeds and woods.

Pharmacological activities

Nutmeg and mace, widely accepted as flavoring agents, are used in higher doses for their aphrodisiac and psychoactive properties and antimicrobial activity.

Chemical constituents

Nutmeg seeds contain fixed oil, commonly called nutmeg butter. This oil contains myristic acid, trimyristin, and glycerides of lauric, tridecanoic, stearic, and palmitic acids. The essential oil contains myristicin, elemicin, eugenol, and safrole. Also present in the oil are sabinene, cymene alpha-thujene, gamma-terpinene, and monoterpene alcohols in smaller amounts. Phenolic compounds found in nutmeg are reported to have antioxidant properties. Other isolated compounds include the resorcinol smalabaricone B and malabaricone C, as well as lignans and neolignans.

13. *Ocimum basilicum* Linn. (Lamiaceae)

Common names
Eng.: Basil, Sweet basil.
Hindi & Beng.: Babui, Tulsi, Sabzah and Babui-tulsi

Distribution

It is indigenous to Persia and Sind, is cultivated in gardens in India.

Parts used: Leaves

Pharmacological activities

Carminative, stimulant, diaphoretic, diuretic and antimicrobial.

Chemical constituents

Essential oils (linalool, estragol and eugenol); tannins and flavonoids.

14. *Origanum vulgare* Linn. (Labiatae)

Common names
Eng.: Mediterranean oregano, mountain mint, wild marjoram, winter marjoram, winter sweet.
Hindi & Beng.: Sathra, Mridu-maru-vama.

Distribution

Common or wild oregano is a perennial plant native to the Mediterranean region and Asia and cultivated in the United States. Its creeping root stock produces a square, downy, purplish stem with opposite ovate leaves.

Parts used: Leaves

Pharmacological activities

Aside from its culinary application, oregano exhibits antimicrobial and antioxidant actions and has possible activity as an antispasmodic and in diabetes. However, there is no clinical evidence to support the use of oregano in any indication.

Chemical constituents

Oregano contains oleanolic and ursolic acids, flavonoids and hydroquinones, caffeic, rosemarinic, and lithospermic acid, tannins, and phenolic glycosides. Phenolic compounds represent 71% of the total oil. The polar phenols thymol and carvacrol are responsible for many of the properties of the essential oil, as well as p-cymene and terpinene.

15. *Pimpinella anisum* Linn. (Umbelliferae)

Common names
Eng.: Anise, aniseed, sweet cumin.
Hindi & Beng.: Saonf, Saurif, Sonf and Muhuri, Mithi-jira.

Distribution

It has been cultivated in Egypt for at least 4,000 years. Records of its use as a diuretic and treatment of digestive problems and toothache are seen in medical texts from this era.

Parts used: Fruits.

Pharmacological activities

Anise is used as a flavoring in alcohols, liqueurs, dairy products, gelatins, puddings, meats, and candies, and as a scent in perfumes, soaps, and sachets. The oil has also been used to treat lice, scabies, and psoriasis. Anise frequently is used as a carminative and expectorant. Anise also is used to decrease bloating and settle the digestive tract in children. In high doses, it is used as an antispasmodic and an antiseptic and for the treatment of cough, asthma, and bronchitis.

Chemical constituents

Anise oil (1% to 4%) is obtained by steam distillation of the dried fruits of the herb. The highest quality oils result from anise seeds of ripe umbels in the center of the plant. A major component of the oil is trans-anethole (75% to 90%), responsible for the characteristic taste and smell, as well as for its medicinal properties.

The volatile oil also has related compounds that include estragole (methyl chavicol, 1% to 2%), anise ketone (p-methoxyphenylacetone), and beta caryophyllene. In smaller amounts are anisaldehyde, anisic acid, limonene, alpha-pinene, acetaldehyde, p-cresol, cresol, and myristicin (the psychotomimetic compound previously isolated from nutmeg).

Constituents of the whole seed include coumarins, such as umbelliferone, umbelliprenine, bergapten, and scopoletin. Lipids (16%) include fatty acids, beta-amyrin, stigmasterol, and its salts. Flavonoids in aniseed include rutin, isoorientin, and isovitexin. Protein (18%) and carbohydrate (50%) are also present. Terpene hydrocarbons in the plant also have been described.

16. *Psidium guajava* Linn. (Myrtaceae)

Common names

Eng.: Guava, Goiaba, Guayaba, Djamboe
Hindi & Beng.: Lalsufrium and Lalpeyara; Goachiphal.

Distribution

Southern Mexico into or through Central America. It is common throughout all warm areas of tropical America and in the West Indies.

Parts used: Leaves

Pharmacological activities

Cure for diarrhea, Indians also employ it for sore throats, vomiting, stomach upsets, for vertigo, to regulate menstrual periods and antibacterial.

Chemical constituents

Guava is rich in tannins, phenols, triterpenes, flavonoids, essential oils, saponins, carotenoids, lectins, vitamins, fiber and fatty acids. Coumarins, essential oils, ellagitannins.

17. *Punica granatum* Linn. (Punicaceae or Lythraceae)

Common names

Eng.: Pomegranate
Hindi & Beng.: Anar; Dhalim and Dalimb

Distribution

The pomegranate is considered to have originated in the region from Iran to northern India, and has been cultivated since ancient times throughout the Mediterranean region. Today, it is widely cultivated throughout the Middle East and Caucasus region, northern Africa and tropical Africa, the Indian subcontinent, Central Asia, and the drier parts of Southeast Asia. It is also cultivated in parts of California and Arizona. In recent years, it has become more common in the commercial markets of Europe and the Western Hemisphere.

Parts used: Pericarp.

Pharmacological activities

Fever, malaria, febrifuge and antimicrobial.

Chemical constituents

Ellagitannins, tannin, pectin and alkaloids.

18. *Rosmarinus officinalis* Linn. (Lamiaceae)

Common names
Eng.: Rosemary, anthos.
Hindi & Beng.: Rusmari.

Distribution

It is native to the Mediterranean region.

Parts used: Leaves

Pharmacological activities

Carminative, antimicrobial and stimulant.

Chemical Constituents

Flavonoids, phenolic acids (caffeic, chlorogenic and rosmarinic) and essential oils (camphor and cineole) and diterpenes (carnosol).

19. *Salvia officinalis* Linn. (Lamiaceae)

Common names
Eng.: Sage, Garden sage
Hindi & Beng.: Salbia-sefakuss.

Distribution

Native to the Mediterranean region, though it has naturalized in many places throughout the world.

Parts used: Leaves

Pharmacological activities

Stimulant, astringent, tonic and carminative, antimicrobial and dyspepsia.

Chemical constituents

Rosmarinic, caffeic, chlorogenic acids, carnosol, flavonoids, essential oils (mainly thujone and cineole).

20. *Sambucus nigra* Linn. (Caprifoliaceae)

Common names
Eng.: Elderberry, Arizona elderberry, American elder, sweet elder, wild elder, flor sauco, tree of music, danewort, walewort, new Mexican elderberry, velvet-leaf elder, hairy blue elderberry, and dwarf elder.

Distribution

Common elderberry is common along stream banks, river banks, and open places in riparian areas lower than < 3000 m. From west Texas north to Montana, western Alberta, and southern British Columbia, and all other western states, south into Northwest Mexico.

Parts used: Leaves, roots, seeds, and berries.

Pharmacological activities

Elderberry is used as a remedy for viral infections like the flu and common cold. Elder stimulates the circulation, causing sweating, effectively cleansing the body.

Chemical constituents

The active alkaloids in elderberry plants are hydrocyanic acid and sambucine, cyanide, cyanogenic glucoside and sambunigrin.

21. *Sassafras albidum* Nees. (Lauraceae)

Common names
Eng.: Sassafras, saxifras, ague tree, cinnamon wood, saloop.

Distribution

Native to eastern Asia and one native to eastern North America. Fossils show that sassafras was once widespread in Europe, North America and Greenland.

Parts used: All party including stern, leaves, bark, wood, roots, fruits and flowers.

Pharmacological Activities

Antimicrobial and antifungal activity, other external uses include treatment of rheumatism, gout, sprains, swelling and cutaneous eruptions. A recent report compares safrole (the main constituent from sassafras oil), to indomethacin for anti-inflammatory activity and pain treatment in mice.

Chemical constituents

The main constituent of sassafras oil is safrole, which chemically is p-allyl-methylene dioxybenzene, which comprises up to 80% of the oil. Volatile oil also contains anethole, pinene apiole, camphor, eugenol and myristicin. The plant contains less than 0.2% total alkaloids (primarily boldine and its derivatives and reticuline) along with tannins, resins, mucilage and wax. Six alkaloids, aporphine and benzylisoquinoline derivatives, have been found in root bark.

22. *Sida rhombifolia* Linn. (Malvaceae)

Common names
Eng.: Country mallow, Yellow barleria.
Hindi & Beng.: Lalbariala

Distribution

It is found all over India.

Parts used: Root, stem and leaves.

Pharmacological Activities

Leaves and roots used, piles, gonorrhea, anti-sound, diuretic, aphrodisiac. It is also used in the treatment of rheumatism and anti-bacterial activity. Stems used as emollient and demulcent.

Chemical Constituents

Ephedrine and pseudoephedrine found in the leaves.

23. *Syzygium aromaticum* Linn. (Myrtaceae)

Common names
Eng.: Clove, caryophyllus
Hindi & Beng.: Laung

Distribution

The clove plant grows in warm climates and is cultivated commercially in Tanzania,

Sumatra, the Maluku (Molucca) Islands, and South America.

Parts used: Dried buds

Pharmacological activities

Clove is typically used as a topical analgesic (it has a natural numbing quality), it can also be used to combat intestinal bacteria. For internal use, use in cooking or make a cup of clove tea.

Chemical constituents

Essential oils (eugenol), taninns and flavonoids.

24. *Syzygium cumini* Skeels. (Myrtaceae)

Common names
Eng.: Jambolan, Java plum, Jambul, Jambolan, Jamblang.
Hindi & Beng.: Jamun; Jambhul

Distribution

S. cumini is a fast-growing tropical and subtropical tree preferring moist, riverine habitats, that is valued for its fruit, timber and as ornamental and as such has been widely introduced from its native South Asia.

Parts used: Leaves.

Pharmacological Activities

S. cumini fruit have a sweet or sub-acid flavor with little astringency and antimicrobial.

Chemical Constituents

Flavonoids and tannins.

25. *Thymus vulgaris* Linn. (Lamiaceae)

Common names
Eng.: Garden Thyme, Rubbed Thyme.
Hindi & Beng.: Thyme

Distribution

The plant is indigenous to the Mediterranean and neighboring countries, northern Africa, and parts of Asia. It is extensively cultivated.

Parts used: Leaves and flowers

Pharmacological activities

Thyme is used for cough and bronchitis as it produces expectoration and reduces bronchial spasm. It also makes a good use in indigestion, gastritis, antimicrobial and diarrhea.

Chemical Constituents

Essential oils (mainly thymol and carvacrol), flavonoids, tannins and triterpenes.

26. *Tribulus terrestris* Linn. (Zygophyllaceae)

Common names
Eng.: Small Caltrops.
Hindi & Beng.: Chota-gokhru and Gokhuri

Distribution

It is found all over India and Ceylon.

Parts used: Root and fruit.

Pharmacological Activities

Antifungal activity, cooling, diuretic and demulcent.

Chemical Constituents

Saponins, resin and alkaloids.

27. *Uncaria tomentosa* Roxb. (Rubiaceae)

Common names
Eng.: Cat's Claw.
Hindi & Beng.:

Distribution

Uncaria tomentosa is a woody vine found in the tropical jungles of South and Central America.

Parts used: Leaves, roots and bark.

Pharmacological Activities

In addition to being an antibacterial, antifungal and antiviral herb, cat's claw is also known for boosting the immune system, increasing your body's protection against illness.

Chemical Constituents

Oxindole alkaloids found in the bark and roots of Cats Claw, which have been documented to stimulate the immune system. It is these seven different alkaloids that are credited with having a variety of different medicinal and healing properties. The most immunologically active alkaloid is believed to be Isopteropodin (Isomer A), which increases the immune response in the body and act as antioxidants to rid the body of free radicals. Compounds found in Cat's Claw may also work to kill viruses, bacteria, and other microorganisms that cause disease, and they work to inhibit healthy cells from becoming cancerous.

28. *Usnea barbata*. (Usneaceae)

Common names

Eng.: Usnea, Usnea Lichen, Usnea Longissima, Beard moss, Tree moss

Distribution

Usnea has a long history of medicinal use throughout this continent, in Europe and in Asia. It was used by many Native American tribes in the Northwestern region of the United States and Canada.

Parts used: The dried thallus or lichen strands of collected plants.

Pharmacological activities

This common lichen is antibacterial and antifungal. A powerful antibiotic, usnea is used to treat urinary tract infections, strep and staph infections, respiratory and sinus infections as well as fungal infections like yeast and vaginosis. It is also used immunostimulant, anti-viral, anti-bacterial, expectorant, anti-Inflammatory, cold and flu.

Chemical Constituents

Lichen acids (usnic acid-antibiotic), polysaccharides (immune stimulating), mucilage, anthraquinones (endocrocin-laxative), fatty acids, all the essential amino acids, vitamins, carotenoids and diffractic acid.

29. *Vaccinium macrocarpon* Aiton. (Ericaceae)

Common names
Eng.: Cranberry, Large cranberry, American cranberry and Bearberry.

Distribution

Vaccinium macrocarpon is native to central and eastern Canada (Ontario to New foundland), and the northeastern and north-central United States (Northeast, Great Lakes Region, and Appalachians as far south as North Carolina and Tennessee). It is also naturalized in parts of Europe and in scattered locations along the Pacific Coast of North America (from California to British Columbia).

Pharmacological Activities

Cranberry is a potent defense against urinary tract infections due to its ability to make the bladder lining too "slippery" to adhere to. Full of antioxidants, cranberry also has antiviral properties and prevents plaque formation on teeth.

30. *Verbascum lasianthum* Boiss. (Scrophulariaceae)

Common names

Eng.: Mullein, velvet plant.
Hindi & Beng.: Gidar-tamaku

Distribution

They are native to Europe and Asia, with the highest species diversity in the Mediterranean.

Parts used: Seeds

Pharmacological activities

Antimicrobial activity against several bacteria and fungi.

Chemical Constituents

It contains mucilage, several saponins, coumarin, triterpenoid and glycosides.

31. *Zingiber officinale* Roscoe. (Zingiberaceae)

Common names

Eng.: Ginger, ginger root, black ginger, zingiberis rhizome.
Hindi & Beng.: Sonth, (fresh) Adrak, Ada, Adi and Ada.

Distribution

A native of tropical Asia, this perennial is cultivated in tropical climates such as Australia, Brazil, China, India, Jamaica, West Africa, and parts of the US. The rhizome is used medicinally and as a culinary spice.

Parts used: Rhizomes.

Pharmacological activities

Ginger and its constituents have antiemetic, cardiotonic, antithrombotic, antibacterial, antioxidant, antitussive, antihepatotoxic, anti-inflammatory, antimutagenic, stimulant, diaphoretic, diuretic, spasmolytic, immuno stimulant, carminative, and cholagogue actions. Ginger is used to promote gastric secretions, increase intestinal peristalsis, lower cholesterol levels, raise blood glucose, and stimulate peripheral circulation. Traditionally used to stimulate digestion, its modern uses include prophylaxis for nausea and vomiting (associated with motion sickness, hypermesis gravidarum, and anesthesia), dyspepsia, lack of appetite, anorexia, colic, bronchitis, and rheumatic complaints. Ginger can be used as a flavoring or spice as well as a fungicide and pesticide.

Chemical constituents

The major constituents in ginger rhizomes are carbohydrates (50 to 70%), which are present as starch. The concentration of lipids is 3 to 8% and includes free fatty acids (*e.g.*, palmitic, oleic, linoleic, linolenic, capric, lauric, myristic), triglycerides, and lecithins. Oleoresin provides 4 to 7.5% of pungent substances as gingerol homologues, shogaol homologues, zingerone, and volatile oils. Volatile oils are present in 1 to 3% concentrations and consist mainly of the sesquiterpenes beta-bisabolene and zingiberene; other sesquiterpenes include zingiberol and zingiberenol; numerous monoterpenes are also found. Amino acids, raw fiber, ash, protein, phytosterols, vitamins (*i.e.*, nicotinic acid and vitamin A), and minerals are among the other

10

MEDICINAL PLANTS FOR DIURETIC

INTRODUCTION

Diuretics or "water pills" are medicines that aid in the elimination of sodium (salt) and water from the body. They act by increasing the excretion of sodium in the urine. When the kidneys excrete sodium, they excrete water along with it. This decreases the blood volume and reduces pressure of the blood on the walls of the arteries. Diuretics are used to treat several conditions, such as high blood pressure, heart failure, liver disease and certain types of kidney disease.

Diuretics are among the most commonly prescribed medications. They inhibit electrolyte reabsorption from the lumen of the nephron, increasing osmolarity and enhancing water and electrolyte excretion. It is important to note that there is a delicate balance between dietary sodium intake and sodium loss. If the balance is compromised and there is a greater intake of sodium into the body, but not enough sodium removal, complications of fluid overload occur, such as oedema, pulmonary oedema or high blood pressure. When there is greater removal of sodium, but not enough sodium intake, complications of fluid depletion take place, such as renal failure or reduced output of blood from the heart. The ability to induce a negative fluid balance has made diuretics useful in the treatment of a variety of conditions, particularly in hypertension and oedematous states.

Renal Handling of Sodium and Water

To understand the action of diuretics, it is important to review the general mechanism by which sodium is reabsorbed and how the kidney filters fluid and forms urine provides an overview of how the kidneys regulate and maintain water and electrolyte balance in the body.

Management

Diuretics cause an increase in urine output. It is important to note that, except for the osmotic diuretics, these drugs typically enhance the excretion of solute and water. Therefore, the net effect of most diuretics is to decrease plasma volume. Most diuretics produce diuresis by inhibiting the reabsorption of sodium at different segments of the renal tubular system. Sometimes a combination of two diuretics is given because this can be significantly more effective than either compound alone, *e.g.*, a synergistic effect. The reason for this is that one nephron segment can compensate for altered sodium reabsorption at another nephron segment.

Therefore, blocking multiple nephron sites significantly enhances efficacy. Diuretics have different clinical uses depending on their mechanism of action.

Five classes of diuretics and their major sites of action include:

1. Thiazide diuretics: Distal tubule
2. Loop diuretics: Ascending limb of the Loop of Henle
3. Potassium-sparing diuretics: Cortical-collecting duct
4. Osmotic diuretics
5. Carbonic anhydrase inhibitors

Thiazide or Low-Ceiling Diuretics

Thiazide diuretics are the most commonly used. The thiazide diuretics are sulphonamide derivatives which act through the inhibition of sodium and chloride reabsorption at proximal sites of the distal kidney tubules. Because this sodium-chloride transporter normally only reabsorbs approximately 5% of filtered sodium, these diuretics are less efficacious than loop diuretics in producing diuresis and natriuresis. Nevertheless, they are sufficiently powerful to satisfy most therapeutic needs when a diuretic is required. Their mechanism depends on renal prostaglandin production. The excretion of potassium, magnesium and zinc is also enhanced, while calcium excretion is diminished. Thiazides, *e.g.*, hydrochlorothiazide, and the thiazide-like diuretic, indapamide, are used mainly in low doses in the treatment of hypertension, for the management of oedema associated with nephrotic syndrome, liver cirrhosis, heart failure, idiopathic hypercalciuria and nephrogenic diabetes insipidus. High doses are not recommended because of biochemical repercussions, such as hypokalaemia.

Loop or High-Ceiling Diuretics

Loop diuretics, *e.g.*, furosemide, bumetanide, torasemide and piretanide, are widely used for the symptomatic treatment of heart failure and fluid retention in chronic kidney disease. Loop diuretics act primarily by inhibiting chloride and sodium reabsorption over the entire length of the thick ascending limb of the loop of Henle. The sodium-potassium-chloride co-transporter in the thick ascending limb normally reabsorbs approximately 25% of the sodium load. Inhibition of this pump leads to a significant increase in the distal tubular concentration of sodium, reduced hypertonicity of the surrounding interstitium and less water reabsorption in the collecting duct. This altered handling of sodium and water leads to both diuresis and increased loss in sodium. By acting on the thick ascending limb, which handles a significant fraction of sodium reabsorption, loop diuretics are powerful. These drugs also induce the renal synthesis of prostaglandins and this contributes to their renal action, including an increase in renal blood flow and the redistribution of renal cortical blood flow. The effect of the loop diuretics is dose dependent, and is largely determined by the rate at which the diuretic is delivered to its site of action. Diuresis is not seen with very low doses.

Progressively increasing diuresis is achieved at higher doses. A plateau is reached, at which even higher doses produce no further dieresis. This dose is called the maximum effective dose, which increases with worsening renal function. Furosemide is chemically similar to the thiazide diuretics. It has a fast onset of diuretic action. The oral form is used for hypertension, either alone or in combination with other anti-hypertensive agents, but the thiazide-type diuretics are preferred, unless there is renal impairment or cardiac failure. The intravenous route is fast acting in emergency situations, such as pulmonary oedema. Furosemide has an acute haemodynamic effect in such circumstances.

Loop diuretics also have an important effect on renal calcium handling. The reabsorption of calcium in the loop of Henle is primarily passive, being driven by the electro chemical gradient created by NaCl transport. As a result, inhibiting the reabsorption of NaCl leads to a parallel reduction in that of calcium, thereby increasing calcium excretion. A concern is that the calciuric response can lead to kidney stones and/or nephro calcinosis. (Nephrocalcinosis is a disorder in which too much calcium is deposited in the kidneys).

Potassium-Sparing Diuretics

Unlike loop and thiazide diuretics, some potassium-sparing diuretics do not act directly on sodium transport. They also antagonise the actions of aldosterone (aldosterone receptor antagonists, *e.g.*, spironolactone) at the distal segment of the distal tubule. This causes more sodium (and water) to pass into the collecting duct and to be excreted in the urine. They are called potassium-sparing diuretics because they do not produce hypokalemia, like the loop and thiazide diuretics. The reason for this is that by inhibiting aldosterone-sensitive sodium reabsorption, less potassium and hydrogen ions are exchanged for sodium by this transporter, and therefore less potassium and hydrogen are lost to the urine. Spironolactone is a competitive inhibitor of aldosterone. It is an effective anti-hypertensive, which may be particularly useful in patients with resistant hypertension. It has the advantage of not affecting glucose metabolism or uric acid elimination. It is the drug of choice for hypertension due to primary aldosteronism. Eplerenone is an important alternative to spironolactone in patients experiencing oestrogenic side-effects. Spironolactone and eplerenone are used in the treatment of hypertension, heart failure and oedema in liver failure. Spironolactone and eplerenone are also known as aldosterone antagonists. Amiloride and triamterene inhibit sodium reabsorption in the distal tubule. They are weak diuretics. When administered alone.

Amiloride also has some antihypertensive activity. These agents should be used with caution in patients at risk of hyperkalemia, such as in patients with renal impairment and those taking angiotensin-converting enzyme inhibitors or potassium supplements.

Carbonic Anhydrase Inhibitors

Carbonic anhydrase inhibitors inhibit the transportation of bicarbonate out of the proximal convoluted tubule into the interstitium, which leads to less sodium reabsorption at this site, and therefore greater sodium, bicarbonate and water loss in the urine. These are the weakest of the diuretics, and are seldom used in cardiovascular disease. Their main use is in the treatment of glaucoma. Carbonic anhydrase inhibitors, such as acetazolamide, are used for the prophylaxis of mountain sickness, which is an unlicensed indication.

Osmotic Diuretics

Osmotic diuretics, *e.g.*, mannitol, are used in a hospital setting for the treatment of cerebral oedema. Mannitol is a non-reabsorbable sugar alcohol that acts as an osmotic diuretic, inhibiting sodium and water reabsorption in the proximal tubule, and more importantly, the loop of Henle. As with loop diuretics, mannitol produces relative water dieresis in which water is lost in excess of sodium and potassium. Mannitol is not generally used in oedematous states since initial retention of the hypertonic mannitol can induce further volume expansion, which precipitates pulmonary oedema in heart failure.

Side-Effects

The most common side-effect associated with diuretics is the increased elimination of potassium, resulting in a dangerously low level of potassium in the body. With the exception of potassium-sparing diuretics, diuretics cause loss of potassium, which, if left untreated, can increase the risk of serious heart rhythm disturbances. Taking a potassium supplement or eating high-potassium foods, such as bananas and orange juice, helps to maintain healthy potassium levels. A potential side-effect of potassium-sparing diuretics is a dangerously high level of potassium in people who already have a high potassium level or in those who have kidney disease. Diuretics cause the patient to pass more urine than usual, and therefore should not be taken at night before bedtime. Increased urination occurs most frequently in people taking loop diuretics. This side-effect improves within a couple of weeks of taking the diuretic in most people. When taking diuretics, the patient will need to have regular check-ups to monitor potassium levels and kidney function. Other potential side-effects of diuretics include:

- Increased thirst
- Urinary incontinence
- Heart palpitations
- Muscle cramps, fatigue or weakness from low potassium levels
- Dizziness or light-headedness
- Numbness or tingling sensations
- Headaches
- A rash
- Impotence
- Menstrual irregularities
- Depression
- Irritability

Allergic reactions: Thiazide diuretics should not be used if the patient is allergic to sulphonamides.

PLANTS FOR DIURETIC

1. *Abelmoschus esculentus* Linn. (Malvaceae)

Common names
Eng.: Edible Hibiscus, Ladies fingers.
Hindi & Beng.: Bendi, Bhindi and Dheras.

Distribution

It is found throughout the India

Parts used: Ripe seeds or unripe fruit.

Pharmacological activities

Emollient, demulcent, diuretic, cooling and aphrodisiac.

Chemical constituents

Albuminoids, carbohydrates, pectin and starch.

2. *Abutilon graveolens* Linn. (Malvaceae)

Common names
Hindi & Beng.: Barkanghi

Distribution

Throughout the tropical parts of India.

Parts used: Whole part

Pharmacological activities

Diuretic and mucilaginous

Chemical constituents

Asparagin.

3. *Abutilon indicum* **G. Don. (Malvaceae)**

Common names
Eng.: Country-mallow
Hindi & Beng.: Kangahi; Kanghani and Potaree.

Distribution

Throughout the tropical parts of India and Ceylon.

Parts used: Root, bark, leaves, seeds and fruit.

Pharmacological activities

Leaves are demulcent, diuretic, laxative, pulmonary, sedative and aphrodisiac. Bark is astringent and diuretic. Root is diuretic; seeds are laxative, expectorant and demulcent.

Chemical constituents

Leaves contain mucilage, tannin, asparagin, organic acid, traces and alkaline, sulphates, chloride, magnesium, phosphate and calcium carbonate. Roots also contain asparagin.

4. *Achras sapota* **Linn. (Sapindaceae)**

Common names
Eng.: Sapodilla plum
Hindi & Beng.: Sapota

Distribution

Native of America, a small tree of slow growth, cultivated throughout the Bombay Presidency, thriving best near the sea.

Parts used: Fruit when ripe are delicious and are eaten.

Pharmacological activities

Diuretic, febrifuge and tonic.

Chemical constituents

Glucoside, alkaloid and sapotin.

5. *Achyranthes bidentata* **Linn. (Amaranthaceae)**

Common names
Eng.: Ox Knee
Hindi & Beng.: Putkanda

Distribution

Temperate and subtropical Himalayas from Kishtwar to Sikkim

Parts used: Seeds and root

Pharmacological activities

Antimicrobial and diuretic.

Chemical constituents

Oligosaccharide, steroids, triterpenoids, alkaloids and coumarins.

6. *Aconitum ferox* Wall. (Ranunculaceae)

Common names
Eng.: Indian aconite; Monkshood
Hindi & Beng.: Mithazahar and Katbish or Mithavish

Distribution

Eastern temperate and sub-Alpine regions of the Himalayas, eastward of Kumaon, Nepal, Kashmir and Sikkim.

Parts used: Dried tuberous root

Pharmacological activities

Diaphoretic, diuretic, antiperiodic, anodyne, antidiabetic, antiphlogistic, antipyretic in very small doses. In large dose it is virulent poison, narcotic and powerful sedative.

Chemical constituents

Toxic alkaloid called Nepelline or pseudo aconitine similar to aconitine also consist of picro- aconine, aconine, benzylaconine and homo-napelline.

7. *Aerva lanata* (Amaranthaceae)

Common names
Eng.: Mountain Knot Grass
Hindi & Beng.: Chhaya and chaya

Distribution

Tropical parts of India.

Parts used: Entire plant

Pharmacological activities

Demulcent, anthelmintic, diuretic and antidiarrhoeal. This herb is described as one of the best known remedies for bladder and kidney stones. Ayurvedic practitioners recommend a decoction of the plant to be taken internally for a few days to dissolves the stone and to clear the urinary path.

Chemical constituents

Palmitic acid, β-sitosterol, alpha-amyrin and alkaloids.

8. *Adansonia digitata* Linn. (Malvaceae)

Common names
Eng.: Boabab or monkey-bread tree of Africa.
Hindi & Beng.: Gorakhamli, Sumpura.

Distribution

One of the largest and long lived trees in the world, met with chiefly in Bombay, Gujarat and Coromandel Coast and Ceylon.

Parts used: Pulp of fruit, bark and leaves.

Pharmacological activities

Fruit is somewhat acid, refrigerant and diuretic. Seed and its pulp are astringent, demulcent, stomachic and anti-scorbutic.

Chemical constituents

Pulp contains phlobaphenes, mucilage and gum, glucose, tartrate and acetate of potash and other salts.

9. *Adiantum capillus veneris* Linn. & Bedd. (Polypodiaceae)

Common names
Eng.: Maiden-hair fern.
Hindi & Beng.: Hansraj, Mubaraka.

Distribution

Chiefly obtained in the Punjab bazaars and in some parts of Southern India.

Parts used: Leaves and rhizomes

Pharmacological activities

Expectorant, diuretic and emmenagogue. Syrup prepared from the leaves is useful in chronic cough. This plant is also used as antidandruff, antitussive, demulcent, depurative, emetic, galactagogue, laxative, stimulant and tonic.

Chemical constituents

Triterpenes, flavonoids, phenylpropanoids and carotenoids.

10. Aerva lanata Juss. (Amaranthaceae)

Common names

Eng.: Astmabayda.
Hindi & Beng.: Chaya.

Distribution

It is common weed in South India.

Parts used: Entire plant

Pharmacological activities

Anthelmintic and diuretic.

Chemical constituents

Palmitic acid β-sitosterol, alpha-amyrin and alkaloids.

11. *Agaricus albus* (Fungi)

Common names

Eng.: White Agaric, touchwood
Hindi & Beng.: Chhatri

Distribution

Punjab, Asia Minor.

Parts used: Fungus of the larch, quercus and *Fagus* species.

Pharmacological activities

Astringent, cathartic and lactifuge, diuretic and expectorant.

Chemical constituents

Resin, bitter extractive matter, gum, vegetable albumen and wax. The true active principles are agaric and fungic or larcic acid and also contain Agaricin.

12. *Agave americana* Linn. (Amaryllidaceae)

Common names

Eng.: American aloe; Carate.
Hindi & Beng.: Bara-khawar and Mah-Jangli-anarash; Ghayal.

Distribution

It is found in many parts of India.

Parts used: Roots, leaves and gum.

Pharmacological activities

Diuretic and anti-syphilitic. Sap is laxative, diuretic, emmenagogue and antiscorbutic.

Chemical constituents

Agavose is an inactive sugar and saponins.

13. *Agrimonia eupatorium* Linn. (Rosaceae)

Common names
Eng.: Common Agrimony, Cockeburr, Sticklewort, Philanthropos.
Hindi & Beng.: Church Steeples

Distribution

The plant is found abundantly throughout England, on hedge-banks and the sides of fields, in dry thickets and on all waste places. In Scotland it is much more local and does not penetrate very far northward.

Parts used: Herb

Pharmacological activities

Aromatic, astringent, anthelmintic and diuretic. It is useful agent in skin eruptions and diseases of the blood, pimples, blotches, etc.

Chemical constituents

Essential oil

14. *Ajuga bracteosa* Wall. (Labiatae)

Common names

Eng.: Neelkanthi, Leelkounthe.

Hindi & Beng.: Ratpacho, Lilkounthe.

Distribution

It is cultivated in Jammu-Kashmir Himalaya.

Parts used: Leaves

Pharmacological Activities

Bitter, astringent, diuretic and aperients.

Chemical Constituents

Glycoside, tannin, palmitic acids, along with glucosidic constituent. Plant also consist of ajugarin I, and lupulin A.

15. *Alhagi camelorum* Fisch. (Papilionaceae)

Common names
Eng.: Khari-i-buz.

Distribution

It is found throughout in India

Parts used: Herb

Pharmacological activities

Plant laxative, diuretic, antibilious and antiseptic. Decoction of roots used for swellings and abscesses. Sugary exudation obtained from the plant used as an expectorant, anti-emetic and laxative.

Chemical constituents

Beta-phenethylamine, N-methyl-Beta-phenethylamine, N-methyltyramine, hordenine, 3, 4-dihydroxy-Beta-phenethyl-trimethyl ammonium hydrozide, 3-methoxy-4-hydrozyl-Beta-phenethyl trimethy lammoniun hydroxide, N-methyl-mescaline, salsolidine, small quantity of choline and betaine. Roots and stems contain alkaloid, also phytosterols, flavonoids and phenolic compound.

16. *Alhagi maurorum* Desv. (Papilionaceae)

Common names

Hindi & Beng.: Jawasa and Dulal-labha.

Distribution

This shrub is native to the region extending from the Mediterranean to Russia, but has been introduced to many other areas of the world, including Australia, southern Africa, and the western United States.

Parts used: Leaves and herb

Pharmacological activities

Alhagi maurorum has been used locally in folk medicine as a treatment for glandular tumors, nasal polyps, and ailments related to the bile ducts. It is used as a medicinal herb for its gastroprotective, diaphoretic, diuretic, expectorant, laxative, antidiarrhoeal and antiseptic properties, and in the treatment of rheumatism and hemorrhoids

Chemical constituents

Manna.

17. *Allium cepa* Linn. (Liliaceae)

Common names
Eng.: Onion
Hindi & Beng.: Piyaz and Piyaj

Distribution

It is cultivated all over India.

Parts used: Bulb and seeds

Pharmacological activities

Stimulant, expectorant and diuretic.

Chemical constituents

Bulb contain an acid acrid, volatile oil which contains sulphur essential oil and organic sulphides and also consist of quercetin.

18. *Alocasia indica* Schott. (Araceae)

Common names
Eng.: Great-leaved caledium
Hindi & Beng.: Alu, Mankanda and Mankachu.

Distribution

It is cultivated all over India.

Parts used: Roots- stock or tubers, petioles and stems

Pharmacological activities

Degestive, laxative, diurertic, lactagogue and leaves are styptic, astringent, gout, rheumatism and dropsy.

Chemical constituents

Acicular crystals of oxalate of lime to which its acridity is due.

19. *Althaea rosea* Linn. (Malvaceae)

Common names
Eng.: Marsh Mallow/Hollyhock.
Hindi & Beng.: Khatmi

Distribution

It is found throughout the India

Parts used: Leaves, Flower and seeds

Pharmacological activities

Seeds are demulcent, febrifuge and diuretic. Roots are astringent and demulcent.

Chemical constituents

Seeds contain alkaloids, saponins, resins, volatile oils, and calcium, mucilage, starch, phosphate of lime.

20. *Allium sativum* (Liliaceae)

Common names
Eng.: Garlic
Hindi & Beng.: Lashun

Distribution

Native to Central Asia and cultivated throughout India.

Parts used: Bulb

Pharmacological Activities

Antibiotic, bacteriostatic, fungicide, anthelmintic, hypotensive and diuretic.

Chemical Constituents

Sulphur containing amino acids known as alliin.

21. *Anthriscuscere folium* Hoffm. (Umbelliferae)

Common names
Eng.: Atrilal.
Hindi & Beng.: Garden chervil

Distribution

It is found throughout the India

Parts used: Seeds

Pharmacological activities

Diuretic, stomachic and deobstruent.

Chemical constituents

Essential oil, glucoside and apiin.

22. *Azima tetracantha* (Salvadoraceae)

Common names
Eng.: Mulchangan, Needle bush
Hindi & Beng.: Kantagurkamai

Distribution

Peninsular India, Orissa, West Bengal. Widespread from Somalia, through East and Central Africa to Nambia and South Africa. Also in Madagascar, Aldabra, Comoro islands and from Arabia to India and Sri Lanka and the Philippines.

Parts used: Seeds and roots

Pharmacological activities

Stimulant, used in rheumatism and expectorant and diuretic.

Chemical constituents

Alkaloids- azimine, azcarpine, carpine.

23. *Bambusa arundinacea* Retz. (Gramineae)

Common names
Eng.: Bamboo
Hindi & Beng.: Bamboo

Distribution

Common in central and south India, cultivated in Bengal and North-western India.

Parts used: All parts

Pharmacological activities

Leaves are emmenagogue and anthelmintic, stimulant, astringent, febrifuge, tonic, cooling antispasmodic, asphrodisiac and diuretic.

Chemical constituents

Silicum as hydrate of silicic acid, peroxide of iron, potash, lime, aluminia, vegetable matter, cholin, betain, nuclease, urease proteolytic enzyme and cyanogenetic glycosides.

24. *Berberis vulgaris* Linn. (Berberidaceae)

Common names
Eng.: Pipperidge Bush, Berberis Dumetorum.

Distribution

The Common Barberry, a well-known, bushy shrub, with pale-green deciduous leaves, is found in copses and hedges in some parts of England, though a doubtful native in Scotland and Ireland. It is generally distributed over the greater part of Europe, Northern Africa and temperate Asia. As an ornamental shrub, it is fairly common in gardens.

Parts used: Bark, root-bark.

Pharmacological activities

It is used in all cases of jaundice, general debility and biliousness, and for diarrhoea. It is also used as purgative and removing constipation and diuretic.

Chemical constituents

The chief constituent of Barberry bark is Berberine, a yellow crystalline, bitter alkaloid, one of the few that occurs in plants belonging to several different natural orders. Other constituents are oxyacanthine, berbamine, other alkaloidal matter, a little tannin, also wax, resin, fat, albumin, gum and starch.

25. *Blepharisedulis* Pers. (Acanthaceae)

Common names

Hindi & Beng.: Utarjan; Utangan

Distribution

It is found throughout India.

Parts used: Seeds

Pharmacological activities

Resolvent, diuretic, aphrodisiac and expectorant, gastrointestinal, respiratory and inflammatory disorders. It is used in folk medicine to treat asthma, cough, fever, inflammation of throat. It is applied locally to heal fastering wounds and ulcers. It is appetizer, astringent to bowels.

Chemical constituents

Crystalline bitter principle, Allantoin, a bitter glycoside and Blepharin, Its seeds are used as food to increase sperm count and as aphrodisiac plant.

26. *Blepharis maderaspatensis* Linn. (Acanthaceae)

Common names
Hindi & Beng.: Utanjan, Dudhiyachoti, Utangan

Distribution

It is cultivated from the Mysore district, Karnataka, India.

Parts used: Seeds, root and leaves.

Pharmacological activities

Blepharis maderaspatensis is used to treat headache. Seeds are used as dysuria, diseases of nervous system, diuretic, aphrodisiac, it is used to cure cuts and wounds, juice extracted from leaf is heated with gingelly oil and applied on affected places to heal wound. Dry seeds of this plant contain steroids and the plant is used for brain disorders.

Chemical constituents

Flavonoids, phenols, tannins, saponins and cardiac glycosides.

27. *Blumea lacera* Dc. (Compositae)

Common names
Eng.: Bara Kukshima, Kukursunga, Shealmotra, Shealmoti.
Hindi & Beng.: Kukurbenda and Kukursunga; Kulsung

Distribution

The plant occurs throughout the plains of India from the north-west ascending to 2,000 ft in the Himalayas. It is a common roadside weed in Ceylon and Malaya. It is distributed to the Malay Islands, Australia, China and Tropical Africa.

Parts used: Whole plant

Pharmacological activities

Aromatic, astringent, stomachic, antispasmodic, emmenagogue, anti-inflammatory, styptic, ophthalmic, digestive, anthelmintic, liver tonic, expectorant, febrifuge, antipyretic and diuretic.

Chemical constituents

Various parts of the plant yield an essential oil containing cineol, fenchone and Blumea camphor. Leaves also contain coniferyl alcohol derivatives, campesterol and flavones. Ethanolic extract of the aerial parts contain hentriacontane, hentriacontanol, α-amyrin, lupeol and its acetates and β-sitosterol. Root and root bark contain triterpenes and sterols.

28. *Boerhaavia diffusa* Linn. (Nyctagineae)

Common names
Eng.: Spreading hog-weed

Hindi & Beng.: Beshakapore; Gadhaparna; Thikri and Gandhapurna; Swetapoorna.

Distribution

Found all over India.

Parts used: Roots and herb

Pharmacological activities

Expectorant, laxative, bitter, stomachic, diuretic, emetic. Root is purgative, anthelmintic and febrifuge.

Chemical constituents

Alkaloidal in nature (punarnavine).

29. *Benincasa hispida* (Cucurbitaceae)

Common names
Eng.: Kushmanda, Winter melon, Wax gourd, Fuzzy melon, Green pumpkin.
Hindi & Beng.: Petha; Pethakaddu

Distribution

Cultivated largely in Uttar Pradesh, Punjab,

Rajasthan and Bihar.

Parts used: Roots, leaves and fruits

Pharmacological activities

Cooling, treatment of skin bruises and diuretic.

Chemical constituents

Pentacyclic triterpene

30. *Bombaxmalabaricum* Dc. (Malvaceae)

Common names
Eng.: Silk cotton tree
Hindi & Beng.: Nurma, Deokapas, Huttian, Ruktasimal, Shimul.

Distribution

It is found throughout the hotter forest regions of India. Cultivated also in gardens.

Parts used: Gum, seeds, leaves, fruit or capsule, bark, flower.

Pharmacological activities

Astringent, demulcent, diuretic and tonic. Roots are emetic. Flower are laxative and diuretic.

Chemical constituents

Gallic acid, tannic acid and gum.

31. *Borassus flabellifer* Linn. (Palmae)

Common names
Eng.: Ice-apple
Hindi & Beng.: Taal, Tari and Pana Nangu, Tal

Distribution

The Asian palmyra palm, toddy palm, or sugar palm, is native to the Indian subcontinent and Southeast Asia, including Nepal, India, Bangladesh, Sri Lanka, Cambodia, Laos, Burma, Thailand, Vietnam, Malaysia, Indonesia and the Philippines. It is reportedly naturalized in Pakistan, Socotra, and parts of China.

Parts used: All parts

Pharmacological activities

When pulp from the unripe fruit is diuretic, demulcent and nutritive; terminal buds are nutritive and diuretic. Root is cooling and restorative; juice is diuretic, cooling, stimulant and antiphlogistic.

Chemical constituents

Albuminoids, gum and fats.

32. *Boswellia serrata* Roxb. (Burseraceae)

Common names
Eng.: Boswellia, Indian olibanum
Hindi & Beng.: Luban, Salai

Distribution

Boswellia are moderate-sized flowering plants, including both trees and shrubs, and are native to tropical regions of Africa and Asia. The plant is native to much of India and the Punjab region that extends into Pakistan.

Parts used: Bark, Root

Pharmacological activities

Arthritis, emmenagogue, diaphoretic and diuretic.

Chemical constituents

Essential oil, Boswellic acid and other pentacyclic triterpene acids are present. Beta-boswellic acid is the major constituent.

33. *Brassica alba* Rabenh. (Cruciferae)

Common names
Eng.: White mustard
Hindi & Beng.: Sufedrai and Dhop-rai

Distribution

It is cultivated in India; indigenous to Western Asia.

Parts used: Seeds

Pharmacological activities

Emetic, diuretic and nervine stimulant.

Chemical constituents

34. *Brassica campestris* Linn. (Cruciferae)

Common names
Eng.: Rape
Hindi & Beng.: Shulgam

Distribution

It is found throughout the India.

Parts used: Thick fleshy underground stem or root, tender leaves and seeds.

Pharmacological activities

Cooking, excellent spring medicine and diuretic.

Chemical constituents

Green tops contains potash. Crude rape oil is dark- brown in colour but is refined into a clear yellow oil that possesses a characteristic harsh taste.

35. *Cardiospermum halicacabum* Linn. (Spindaceae)

Common names
Eng.: Balloon vine or winter cherry; heart pea.
Hindi & Beng.: Kanphata and Nayaphataki; Shib-jhul.

Distribution

India, chiefly Bengal and U.P.

Parts used: Herb, roots, leaves and seeds.

Pharmacological activities

Leaves and root are laxative, stomachic, alterative, emetic and diuretic, externally rubefacient.

Chemical constituents

Essential oil, bitter and stimulant and saponin.

36. *Carthamus tinctorius* Linn. (Compositae)

Common names
Eng.: Saf- flower; Wild saffron; Parrot seed
Hindi & Beng.: Kusumbar; Kusum and Kusum; Kajireh

Distribution

Tropical and subtropical parts of India.

Parts used: Root, seeds and flowers.

Pharmacological activities

Root is used as diuretic. Seeds are purgative.

Chemical constituents

Carthamin or carbonite and albuminoids.

37. *Cicer arietinum* Linn. (Papilionaceae)

Common names
Eng.: Chick - common pea
Hindi & Beng.: Chana and gram;

Distribution

A pulse cultivated in Sind, Bombay, presidency and growing wild in southern India.

Parts used: Seeds or peas, leaves

Pharmacological activities

Seeds are astringent and antibilious. Fried seeds are diuretic. Leaves juice are stomachic and laxative.

Chemical constituents

Malic, oxalic and acetic and another acids and albuminoids.

38. *Cichorium intybus* Linn. (Compositae)

Common names
Eng.: Endive, Chicory, Wild chicory
Hindi & Beng.: Hinduba, Kasni

Distribution

North West India, the Deccan, Punjab, Kashmir, Persia and Europe.

Parts used: Seeds, flower and roots.

Pharmacological ativities

Stomachic, tonic and diuretic. Seeds are carminative and cordial. Root is bitter.

Chemical constituents

Bland oil, Potash mucilage, bitter, glucoside, cichoriin, bitter substances, lactucin and intybin.

39. *Cinnamomum iners* Reinw. (Lauraceae)

Common names
Eng.: Cassia cinnamon
Hindi & Beng.: Tejpat (leaves), Dalchini (bark).

Distribution

Tropical and sub-tropical Himalayas, U.P., Eastern Bengal, Khasia and the Jaintia hills and Burma.

Parts used: Bark, oil and leaves

Pharmacological activities

Stimulant, carminative, deobstruent, diuretic, diaphoretic and galactogogue.

Chemical constituents

Essential oil, eugenol, terpene and cinnamic aldehyde.

40. *Cissam pelos pareira* Linn. (Menispermaceae)

Common names
Eng.: Velvet-leaf
Hindi & Beng.: Harjori and Akanadi

Distribution

Tropical and sub-tropical India from Sind and Punjab to South- India and Ceylon.

Parts used: Bark, roots and leaves

Pharmacological activities

Mild stomachic, bitter tonic, diuretic, astringent, sedative action and antilithic.

Chemical constituents

Cissampeline or pelosine in the root sepeerine, beberines and cissampeline.

41. *Citrullus vulgaris* Schrad. (Rutaceae)

Common names
Eng.: Water melon
Hindi & Beng.: Tarbuz; Jamuka and Tarmuj

Distribution

It is cultivated throughout India.

Parts used: Seeds, fruit

Pharmacological activities

Seeds are cooling demulcent, vermifuge, nutritive and diuretic.

Chemical constituents

Seeds contain fixed oil, proteids, citrullin.

42. *Cubeba officinalis* Miq. (Piperaceae)

Common names
Eng.: Tailpepper; cubebs
Hindi & Beng.: Sitalachini, Kababchini

Distribution

Cultivated specially in Mysore state.

Parts used: Dried immature full-grown fruit called the cubebs.

Pharmacological activities

Carminative, stimulant, diuretic and expectorant.

Chemical constituents

Volatile essential oil.

43. *Cuscuta reflexa* (Convolvulaceae)

Common names
Eng.: Dodder
Hindi & Beng.: Akasbel and Algusi, Haldi-algusilata.

Distribution

It is found throughout the India. A parasitic climber common throughout India.

Parts used: Entire plant

Pharmacological activities

Carminative and diuretic.

Chemical constituents

Marbelin and Kaempferol, stem gave, cuscutin, cuscutatin, β-sitosterol, luteolin, bergenin kaempferol, alkaloids

44. *Cucumis anguinus* (Cucurbitaceae)

Common names

Hindi & Beng.: Kakura

Distribution

It is found in Eastern Bengal.

Parts used: Ripe seeds or unripe fruit

Pharmacological activities

Diuretic

Chemical constituents

Serpentine

45. *Cucumis sativus* Linn. (Cucurbitaceae)

Common names
Eng.: Common cucumber
Hindi & Beng.: Kankri and Khira

Distribution

It is cultivated throughout the India.

Parts used: Seeds and leaves.

Pharmacological activities

Fruits are nutrient and demulcent. Seeds are cooling and diuretic.

Chemical constituents

Fixed oil, starch, resin and sugar. Seeds contains much farinaceous matter.

46. *Cucurbita maxima* Duchesne (Cucurbitaceae)

Common names
Eng.: Red gourd, Great pumpkin
Hindi & Beng.: Pilakohola, Kaddu and Saphuri, Kumra

Distribution

It is found throughout the India.

Parts used: Fruit, stalk, seeds and pulp.

Pharmacological activities

Anthelmintic, taenecide and diuretic

Chemical constituents

Saponin, fixed oil, resin, sugar and starch, proteins, and albuminoids.

47. *Curculigo orchioides* Gaertn. (Amaryllidaceae)

Common names
Eng.: Black musale
Hindi & Beng.: Musalikand, kalimusli and Talamuli

Distribution

Hattest region of India and ceylons

Parts used: Tuberous roots and bulbs

Pharmacological activities

Bitter aromatic tonic, demulcent, diuretic and restorative.

Chemical constituents

Resin, tannins, mucilage, fats and starch.

48. *Cyperus rotundus* Linn. (Cyperaceae)

Common names
Eng.: Nut-grass
Hindi & Beng.: Korehi-jhar and Moothoo.

Distribution

It is cultivated specially south India.

Parts used: Tubers or bulbous root.

Pharmacological activities

Stimulant, tonic, demulcent, anthelmintic, stomachic, carminative, diaphoretic, astringent, vermifuge and diuretic.

Chemical constituents

Fat, sugar, gum, carbohydrates, essential oil, albuminous matter.

49. ***Centella asiatica*** **(Umbelliferae)**

Common names
Hindi & Beng.: Kula kudi and Thulkurhi

Distribution

Marshy places throughout

Parts used: Leaves

Pharmacological activities

Sedative, antibiotic

Chemical constituents:

Triterpenoidsaponins

50. ***Centratherum anthelminticum*** **(Asteraceae)**

Common names
Eng.: Iron weed, Ipecac
Hindi & Beng.: Zeraseha, bakshi, buckshi, kalijhiri, kaliziri, somraj, vapchi, janglijeera, ghorajeera, jangli-jeera, ghora-jeera and somraj, kaliziri, hakuch, bakshie, bapchie, babchi

Distribution

This species is globally distributed in Afghanistan, Pakistan, India and Sri Lanka. Within India, it is found throughout on disturbed sites such as roadsides. It is sometimes cultivated in Himalayas and Khasi Hills.

Parts used: Seeds

Pharmacological activities

Hypotensive activity, Anti-diabetic and diuretic.

Chemical constituents

Avenasterol

51. ***Cichorium intybus*** **(Compositae)**

Common names
Eng.: blue daisy, blue dandelion, blue sailors, blue weed, bunk, coffeeweed, cornflower, hendibeh, horseweed, ragged sailors, succory, wild bachelor's buttons, and wild end.

Distribution

It is found throughout the India North West India, Tamil Nadu and parts of Andhra Pradesh.

Parts used: Entire herb

Pharmacological activities

Laxative, cholagogue, mild hepatic and diuretic.

Chemical constituents

Citric and tartaric acids, acetic, lactic, pyruvic, pyromucic, palmitic and tartaric acids.

52. *Cocos nucifera* (Palmae)

Common names
Eng.: Coconut tree
Hindi & Beng.: Coconut

Distribution

Kerala, Tamil Nadu and Karnataka

Parts used: Fruit, husk

Pharmacological activities

Stomachic, laxative and diuretic.

Chemical constituents

Reducing sugars

BIBLIOGRAPHY

A. Albert, British Journal of Expt. Pathol., 1954.

A. Albert, Brit. J. Exptl. Pathol., 1958.

A. H. Bakett, J. Pharm. Pharmacol., 1958.

Barar, F. S. K., In; Essentials of Pharmacotherapeutics, S. Chand and Company Ltd, New Delhi, 4th Edn, 2007.

Brater DC, Mechanism of action of diuretics, [homepage on the Internet]. 2013.

Chatterjee, C. C., In; Human Physiology, 1stEdn, Vol-II, Medical allied Agency, Calcutta, 1998.

Clapp, WL. Renal Anatomy. In: Zhou XJ, Laszik Z, Nadasdy T, D'Agati VD, Silva FG, eds. Silva's Diagnostic Renal Pathology, New York, NY: Cambridge University Press; 2009.

Diuretics, Patient.co.uk [homepage on the Internet], 2014.

Diuretics. Mayo Clinic [homepage on the Internet], 2014.

E. Fischer, R. Muazo, J. Bacteriol., 1947.

E. W. Stearn, A. E. Stearn, J. Bacteriol., 1924.

Fluids and diuretics, MedlinePlus [homepage on the Internet], 2012.

Ganong, W.F., Review of Medical Physiology, Prentice Hall International Inc., 19thedn, 1999.

Ganong, W.F., Review of Medical Physiology, Prentice Hall International Inc., 18thEdn, 1997.

Goodman, L.S., Gilman, A., The Pharmacological Basis of Therapeutics, Macmillan Publishing Co., New York, 10th Edn., 2001.

Gupta M., Mazumder, U.K., Gomathi, P.,Thamil, S.V., Antiinflammatory evaluation of leaves of *Plumeriaacuminata* , BMC *Complementary and Alternative Medicine,*Vol6, 2006.

Gyton and Hall, Textbook of Medical Physiology, Prism Books (PVT) LTD, 9thEdn, 1996.

Heinrich, M., Barnes, J., Gibbon, S., Williamson, E.M., Fundamentals of Pharmacognosy and Phytotherapy, Churchill Livingstone, 2004.

J. W. Churchman, J. Exptl. Med., 1912.

Jeffery, A. G., Charles, A. D., In; Harrison Principle of Internal Medicine, 14thEdn. Vol-I, Mc-Graw Hill health profession division, New York, 1998.

K. D. Tripathi, Essentials of medical pharmacology, 1994.

Kadam, S.S., Mahadic, K.R., Bothara, K.G., Principles of Medicinal Chemistry, NiraliPrakashan, Vol-1, 8thEdn, 2001.

Katzung, B.G., Basic and Clinical Pharmacology, The McGraw-Hill Co., 8thEdn, 2001.

Kavimani S, In; Source of Medicine, NAMAH.Vol. 8(3), 2000.

Khan, I., Khanum, A., Ehanomedicine and Human Welfare, Ukaaz Publication, Vol-3, 1stEdn, 2005.

Klabunde RE, Diuretics. Cardiovascular Pharmacology Concepts [homepage on internet], 2012.

Kokate, C.K., Purohit, A.P., Gokhale, S.B., Pharmacognosy, NiraliPrakashan, Pune, 17thEdn,, 2001.

L. D. Gebbharadt, J. G. Bachtold, Proc. Soc. Exptl. Biol. Med., 1955.

Lingappa, V. R., Farey, K., In; Physiological Medicine A Clinical Approach to Basic Medical Physiology Hayward, Mc-Graw Hill, Medical Publishing Division, USA.

Maton, Anthea; Jean Hopkins, Charles William McLaughlin, Susan Johnson, MaryannaQuon Warner, David LaHart, Jill D. Wright, Human Biology and Health. Englewood Cliffs, New Jersey, USA: Prentice Hall (1993).

Md. Rageeb Md. Usman, Mrs. Surekha D. Salgar, Dr. Mohammed Zuber Shaikh, Appraisal on Nephroprotective Herbal Plants, Unicorn Publication Pvt. Ltd., 2014.

Md. Rageeb Md. Usman, Vaibhav M. Darvhekar, Anti-inflammatory and Anti-pyretic activity of Murrayakoenigiispreng, Unicorn Publication Pvt., 2015.

Minal S. Patil, Md. Rageeb Md. Usman et. al., Staandardization Techniques of Herbal Medicine, Studium Press PVT. Ltd., 2016.

Mohammed Rageeb, Dr. Mohammed Zuber, Traditional System of Herbal Medicine, Lambert Academic Publishing, 2015.

Mohan, H., In; Text book of pathology 4thEdn, Medical publisher (P) Ltd, New Delhi, 2000.

Mrs. Surekha D. Salgar, Md. Rageebet. al., Practical Handbook of Pharmacognosy, Unicorn Publication Pvt. Ltd., 1st Edn.2016.

Murray, R.K., Granner, D.K., Mayes, P. A., Rodwell, V.W., Harper's Biochemistry, Prentice Hall International, USA, 23rdEdn, 1993.

Nadkarni K. M., Indian Materia Medica, Bombay Popular Prakashan, Volume 1st, 1976.

Prabuji, S.K., RAO, G.P., Patil, S.K., Recent Advaces In Medicinal Plants Research, SatishSerial Publishing House, 2005.

Pullaiah, T., Chandrashekhar Naidu, K., Antidiabetic Plants in India and Herbal based Antidiabetic Research, Regency Publication, New Dhelhi, 2003.

Robert Cruickshank, Hand Book of Bacteriology, 1962.

Samanta, M.K., Srinath, Pulok, K.M., and B. Suresh, Herbal Drug Development-A Modern Approach, Indian Journal of Pharmaceutical Education, Vol-33(4), 1999.

Satoskar, R.S., Bhandarkar, S.D., Ainapure, S.S., Pharmacology and Pharmacotherapeutics, Popular Publication, Mumbai, 17thEdn. 200.

Satyvati, G.V., TondonNeeraj, Sharma Madhu, Indigenous Plant Drug for *Diabetes Mellitus*, *Diabetes*Bulleta, 1989.

South African medicines formulary, 11th Edn. Cape Town: Division of Clinical Pharmacology, Faculty of Health Sciences, University of Cape Town; 2014.

T. K. Chatterjee, Herbal Options, Eastern Traders, 1996.

Tortora, G.J., Grabowski, S.R., Principles of Anatomy and Physiology, 8th Edn., Harper Collins College Publishers, 1996.

Tripathi, K. D., In; Essential of Medical Pharmacology, 4thEdn., Jaypee Brothers Medical publisher (P) Ltd, New Delhi, 2001.

Tripathi, K.D., Essentials of Medical Pharmacology, 5thEdn., Jaypee Brothers Medical Publishers (P) Ltd., 1999.

Tripathi, K.D., Essentials of Medical Pharmacology, Jaypee Brothers Medical Publishers (P) Ltd., 5thEdn, 2000.

Vnder, J. A., Sherman, H. J., Luciano, S., D., In; Human Physiology 5thEdn, MC-Graw publishing company, London, 1990.

Wallis, T. E., In; Text Book of Pharmacognosy, CBS Publishers and Distributors, Delhi, 5th edition, 1985.

Williams, D.A., Thomas, L., Lemke, Foye's Principles of Medicinal Chemistry, Lippincott Williams and Wilkins Publication, 5th edition, 2002.

Chandra Y R, (1976), The wealth of India, publication and information directorate.

Chattergi T K, (1997), Herbal options, Calcutta: Eastern Traders.

Handa S S, (1995), Quality control & standardization of herbal raw materials and traditional remedies, *The Eastern Pharmacist*.

Kirtikar K R and Basu B D (2006), Indian medicinal plants. Vol.-II, 2nd edition, fourth Reprint, published by Lalit mohan Basu, Allahabad, India.

Kirtikar K R and Basu B D, (1975), Indian medicinal plants.

Kirtikar K. R, Basu B D, (1980), Indian Medicinal Plants, Volume 2, Dehradun: International Book Distributors.

Mukherjee P K, (2002), Quality control of herbal drugs, 1st edition, Business horizons pharmaceutical publishers, New Delhi.

Nandkarni K M, (1976), The Indian materia medica, volume 1, Mumbai: popular - prakashan private limited.

Prakash Ghadi, (2001), Pathophysiology for Pharmacy, Second edition, Career Publication, Nashik.

Satyavati G V, Tandon Neeraj, Sharma Madhu, (1989), Indigenous plant drugs for diabetes mellitus, *Indian Journal Diabetes in Developing Countries*.

Savitri Ramaiah, Diabetes, an essential and informative guide to diabetes, Prakash Ghadi.

Savitri Ramajah, Diabetes, (2004), an essential and informative to diabetes, reprint, published and printed by Sterling publishers Pvt. Ltd., New Dehli-110 020.

Schuppan D, Jia J D, Brinkhaus B, (1999), Herbal product for liver diseases: A therapeutic challenge for the new millennium, *Hepatology*.

Sembulingum K, Prema Sembulingum, (2003), Essential of medical physiology, 2nd edition, published by JAYPEE Brother medical publishers (P) Ltd., New Delhi, 110 002, India.

The Wealth of India, (1950), A Dictionary of Indian Raw Materials and Industrial Products. New Delhi: CSIR.

Tripathi K D, (2004), Essential of Medicinal pharmacology, 5th edition, published by JAYPEE Brother medical publishers (p) Ltd., New Delhi 110 002, India.

Tyler V E, (1987), The new honest herbal, G.F. Stickley Co., Philadelphia. USA.

Tyler V E, (1999), Phytomedicine: back to the future, *Journal of natural product*, 62.

WHO (2002), WHO monographs on selected medicinal plants, World Health Organization, Geneva.

WHO (2005), Global atlas of traditional, complementary and alternative medicine, World Health Organization, Geneva.

SUBJECT INDEX

A

B

C

D

E

F

G

P

Q

R

T

U

V

W

X

Z

❑❑❑